DATE DUE

		OCT '90	

DEMCO 25-380

VGM Opportunities Series

OPPORTUNITIES IN
PETROLEUM CAREERS

Gretchen Dewailly
Krueger

Foreword by
Allen E. Murray
Chairman, President, and Chief Executive Officer
Mobil Corporation

VGM Career Horizons
a division of *NTC Publishing Group*
Lincolnwood, Illinois USA

Cover Photo Credits:
Front cover: upper left, Gulf Oil Services photo;
upper right, Texaco USA photo; lower left, photo by
Dennis Sullivan, Lafayette, Louisiana; lower right,
American Petroleum Institute photo.
Back cover: upper and lower left, American
Petroleum Institute photos; upper right, Texaco
USA photo; lower right, photo by Dennis Sullivan,
Lafayette, Louisiana.

Library of Congress Cataloging-in-Publication Data

Krueger, Gretchen Dewailly, 1952–
    Opportunities in petroleum careers / Gretchen Dewailly Krueger ;
    foreword by Allen E. Murray.
        p.   cm. — (VGM opportunities series)
    Bibliography: p.
    ISBN 0-8442-8665-6 : $11.95. — ISBN 0-8442-8666-4 (soft) : $8.95
    1. Petroleum industry and trade—Vocational guidance.   I. Title.
II. Series.
HD9560.5.K78   1989
622'.338'02373—dc20   89-14804
                                                                     CIP

Published by VGM Career Horizons, a division of NTC Publishing Group.
© 1990 by NTC Publishing Group, 4255 West Touhy Avenue,
Lincolnwood (Chicago), Illinois 60646-1975 U.S.A.
All rights reserved. No part of this book may be reproduced, stored
in a retrieval system, or transmitted in any form or by any means,
electronic, mechanical, photocopying, recording or otherwise, without
the prior permission of NTC Publishing Group.
Manufactured in the United States of America.

9 0 VP 9 8 7 6 5 4 3 2 1

ABOUT THE AUTHOR

Gretchen Dewailly Krueger is an English–Journalism graduate of the University of Southwestern Louisiana, the state's second largest institution of higher learning located in Lafayette, Louisiana, the heart of the Gulf of Mexico's oil patch.

Since September 1983, she has been the energy editor at the *Lafayette Daily Advertiser,* where she has won numerous awards for her weekly oil and gas section and other breaking energy-related news stories. Three of the awards are from the Mid-Continent Oil and Gas Association, which selected her tops in the state in the large newspaper category. Two others are second place awards from the Louisiana Press Association for specialized coverage of the petroleum industry and for a special section on LAGCOE (Louisiana Gulf Coast Oil Exposition), Lafayette's biennial oil exposition.

The Bordelonville, Louisiana, native has held public relations positions with several government-affiliated organizations. She is also a member of the Association of Petroleum Writers.

Gretchen Dewailly Krueger is married to Jerry Krueger, a twenty-year veteran of the petroleum industry who she calls her "in-house consultant." They live in Lafayette, Louisiana, and are the proud parents of two cats—Boo and Sweetie.

ACKNOWLEDGMENTS

Although I have made my living as a writer for a number of years, *Opportunities in Petroleum Careers* is my first book. Many hours, phone calls, and volumes of material were used during the research of this vital industry, which touches the daily lives of people throughout the world. However, it could not have been successfully accomplished without the help of so many people and organizations. But even at the risk of leaving someone out, I would like to thank the following:

First, my parents, Mr. and Mrs. V. P. Dewailly, who gave me the opportunity to obtain a college education so I could pursue a writing career. Also, my friend Gen Solar, whose education and experience as a vocational guidance counselor helped me get started as well as kept me on track.

Research material provided by my friend Julie Simon Dronet also made the job so much easier, as did guidance and information from various departments at the University of Southwestern Louisiana, in particular, Dr. Ali Ghalambor, acting head of the Department of Petroleum Engineering. My thanks also to the Petroleum Extension Service at the University of Texas, Austin.

Organizations and associations that were of particular help include the American Petroleum Institute, Washington, D.C.;

Society of Petroleum Engineers, Richardson, Texas; and International Association of Drilling Contractors, Houston, Texas.

Many thanks also go to my good friends John and Frances Love who encouraged me to take on the assignment; Jere Smith, Phillips Petroleum Company's director of media relations, for his help in locating photographs, as well as his wit and words of encouragement; and finally to Brenda Allynn Buras, Texaco USA's Southeast region public and government affairs coordinator, for her time and efforts in getting slides duplicated.

Special thanks for other pictures and assistance go to friends at Digitran and Petroleum Helicopters, Inc., as well as Margaret Badeaux, with Chevron USA; and, my friend Dennis Sullivan, himself a writer, who doubled as my photographer/mentor/critic.

And, last, but certainly not least, my husband Jerry, without whom this book could not have been written.

FOREWORD

If you're looking for excitement, challenges, and opportunities, the petroleum industry offers more than any other business I know. Whether you're thinking of becoming an engineer, an accountant, a geologist, a chemist, a computer programmer, or most anything else, the petroleum business offers careers that may take you further than you ever dreamed—both professionally and geographically.

We work all over the world, and the oil and gas we find and produce keep transportation on the move and factories at work creating products and jobs. Oil and gas also are the raw materials for many plastics and other synthetic materials. Since this is such a vital industry, you know we'll be around long enough for you to build a career.

And for individual companies like Mobil to continue to be successful, we're constantly becoming ever more competitive, efficient, and socially responsible. That means we need not only new technologies, but new talent—employees with the skills, dedication, and ingenuity to meet the energy needs of the twenty-first century. Could that be you?

Consider some of the challenges: we'll need better ways of finding oil and gas in places like the frozen north and the depths of the sea. To get the most out of what we discover, we'll need better drilling and recovery techniques and better producing facilities. We'll need more efficient refineries and petrochemical plants. And to better protect the environment, we'll need safer operations and improved pollution controls. We'll also need to recycle more of our products.

Just as this industry has learned to thrive in hostile natural environments and to meet enormous technological challenges, we've also learned to nurture the necessary human talent for a bright tomorrow. Our industry will continue to be very competitive—but it's as challenging an industry as you'll find. And as the oil business prospers and grows, so will your opportunities.

Allen E. Murray
Chairman, President, and Chief Executive Officer
Mobil Corporation

INTRODUCTION

The discovery of large amounts of petroleum in the middle of the last century catapulted the United States into the Industrial Age and has since left our nation with an unquenchable thirst for one of earth's most precious resources.

Although the industry is suffering through its seventh year of an economic slump, the outlook for career opportunities appears, at this writing, to be excellent for a variety of reasons. First, any industry old-timer will tell you about the roller coaster this business has ridden since it became organized nearly fifty years ago. Many of them have suffered through several boom-to-bust cycles.

Because of America's dependence on petroleum for its many uses, this latest ''bust'' cycle appears to have created the most havoc. It has caused a shortage of personnel in nearly every sector of the industry. Thousands of experienced workers have been lost through layoffs, mergers, and early retirements. Others have simply given up on the cyclical nature of the business and moved into other careers. As early as the summer of 1987, industry watchers were becoming concerned with the lack of qualified workers who had already sought other means of earning a living.

There continues to be a surplus of oil in the world. And, according to the laws of supply and demand, this causes prices to remain low. This in turn makes it uneconomical for U.S. oil companies to drill for new petroleum reserves.

Over the past year, the demand for petroleum and the thousands of products made from it has slowly increased. And, current predictions are that energy prices will rise again in the early 1990s. This is causing some energy specialists to become even more concerned about the already short supply of qualified, trained personnel.

On the other hand, the continuing political turmoil in the Middle East, the ongoing disarray with OPEC, and the current national push for a strong domestic energy policy, is making more Americans aware that a strong U.S. petroleum industry is a must if we are to avoid the oil crises of the 1970s that were highlighted by supply shortages and long gasoline lines.

Providing these trends continue, the petroleum industry should offer a multitude of wide-ranging opportunities for the career-minded student of the 1990s. Anyone, with the proper education and/or training, can have a successful career in the petroleum industry. There are hundreds of careers and opportunities available in any of the major oil and allied companies in the United States and overseas. Many jobs offer on-the-job training, others are of a technical nature, while still others require the completion of a full curriculum at a university or college.

Opportunities in Petroleum Careers gives a detailed explanation of how petroleum is found, taken out of the ground, and made into the products we use each day. It gives details on a full range of career opportunities in today's petroleum industry, as well as the outlook for employment into the next decade. Discussed in the book are salaries and opportunities for advancement.

Also included are an oil field dictionary, a bibliography of sources and recommended reading, appendixes listing where to go for more information, as well as some colleges and universities specializing in petroleum careers.

It would be virtually impossible to include in this book the entire spectrum of this complex, and often technical, industry. However, I believe it is an excellent source of information for both the high school student who is undecided on a future, as well as the adult who may be contemplating a midlife career change.

CONTENTS

What is petroleum and where does it come from? Early uses of petroleum. Drake's Folly. Spindletop. Petroleum in our everyday lives.

Development of the modern-day industry. Takeovers and mergers.

Exploration. Opportunities in petroleum exploration. Leasing lands. Opportunities for landmen/leasing agents.

Drilling onshore. Drilling offshore. Types of rigs. Components of a drilling rig. Safety at the well site. Drilling opportunities. Employment for petroleum engineers.

These petroleum workers probe deep into the ocean floor in search of oil.
(American Petroleum Institute photo)

CHAPTER 1

HISTORY OF PETROLEUM

No other natural resource has helped change the face of the civilized world more than petroleum. In fact, there is probably not one segment of our everyday lives that is not touched in some way by this marvel of nature. From the time we get up in the morning we use petroleum to heat and cool our homes, power the automobile that we drive, and provide raw materials for the thousands of products we use each day. And we are not alone. The lives of millions of people throughout the world are affected by the petroleum industry.

Just over one hundred years ago, the multitude of possibilities petroleum offered could not even be imagined, although the use of petroleum and some of its unrefined products dates back to Biblical times. But, thanks to the persistence and dedication of the early entrepreneurs, who learned how to get it out of the ground, and the inventors, who devised the machines to use it, petroleum has come to play a major role in the everyday lives of each and every one of us.

WHAT IS PETROLEUM AND WHERE
DOES IT COME FROM?

By definition, petroleum, in the strictest sense of the word, is crude oil as it comes out of the ground. It is a mixture of several chemical compounds, primarily hydrogen and carbon. In a broader sense, and for the purpose of this book, petroleum is also defined as all hydrocarbons, including oil, natural gas, natural gas liquids, and all related products. Petroleum can also exist as a solid, such as the tar sands found in some parts of Canada and the oil shale beds located in some western U.S. states.

Even after two hundred years, the origin of black gold, as it is sometimes called, remains the subject of debate. Since the mid-nineteenth century, scientists have variously believed that petroleum comes from coal, decayed animals and vegetables, and even volcanic matter. Today, the debate still continues in some scientific circles as to whether oil is organic (plant or vegetable matter) or inorganic (not living).

The general consensus among the majority of earth scientists, however, is that the petroleum produced today was formed over a period of millions of years when plant and animal matter were compressed as it settled at the bottom of prehistoric seabeds. This matter, covered with layers and layers of sediment, was changed into hydrocarbons through a combination of factors, including bacteria, heat, and pressure.

Oil was first believed to flow under the earth much like an underground stream of water. Further study throughout the years led scientists to learn that oil actually exists between geological structures in areas called reservoirs. The Society of Petroleum Engineers (SPE), one of the industry's largest professional organizations, compares a reservoir rock to a tray of marbles. ''The tray is porous, with open spaces between the grains of the rock, just as there is open space between the marbles,'' according to the SPE. ''When we place the crude oil in the tray, it occupies the pore space in the same way the crude oil occupies pore space in the reservoir rock.''

Oil reservoirs may be a few thousand or many thousands of feet below the surface. How permeable—or how easy the oil or gas flows through connecting pore spaces—determines how easy it will be to remove the petroleum from the ground. How petroleum is found and removed from these reservoirs will be discussed in a later chapter.

EARLY USES OF PETROLEUM

From the beginning of time, humans have learned to be self-sufficient as they designed ingenious methods of utilizing earth's many treasures. The earliest use of petroleum, scientists believe, occurred when natural seepages of both crude oil and natural gas were used by primitive tribes to light their fires. And, according to several versions of the story of the Great Flood, Noah used pitch—a form of natural asphaltic petroleum—as a caulking material to waterproof his Ark.

Indian tribes are said to have used asphalt from the seeps at Santa Barbara, California, as a sealant for their canoes, as well as for warpaints and medicines. And archaeologists believe the ancient Egyptians used the same substance as a lubricant on chariot wheels. Not to be outdone, the Greeks are said to have used petroleum to set the sea on fire to destroy a fleet of ships belonging to an enemy threatening to invade them.

It is said that by the Middle Ages, Sicilians were gathering oil off their coast to use as fuel for lamps. Europeans, in the meantime, skimmed the natural springs for oil that was used for medicinal purposes, as well as for fuel.

It was the inventive Chinese, however, that first drilled for petroleum. Using primitive drilling tools, they bored eight hundred feet into the earth. The year? 347 A.D.—fifteen centuries before the birth of the modern-day petroleum industry!

DRAKE'S FOLLY

By the middle of the nineteenth century, Americans were using oil and gas taken directly from natural deposits on the earth's surface. While they knew these substances burned and could, therefore, be used for light, these pioneers still preferred to use whale oil and tallow candles for heating and lighting purposes. When whale oil began to become more scarce, and, as a result, more expensive, Americans decided to search for an alternative fuel source. These early oil users soon found out, however, that the three-gallon per day production they were getting from naturally occurring oil pools was not nearly enough to meet the growing demands of an expanding nation.

In 1854, Yale University Professor Benjamin Sillman and a group of businesspeople, headed by James M. Townsend, formed what is probably the first oil company in the world—the Pennsylvania Rock Oil Company. This company later became the Seneca Oil Company.

The group had in their possession a report that detailed the suitability of petroleum as a lighting fuel. Based on this report, they decided to attempt the recovery of large amounts of petroleum by taking a cue from water well diggers; they would drill for it.

These early oilpeople hired retired railroad conductor "Colonel" Edwin L. Drake to drill along the banks of Oil Creek at Titusville, Pennsylvania, an area located near an old oil spring. To drill the well, Drake's crew used an old steam engine with an iron bit connected to a rope. This was attached to a wooden winch mechanism they used for hoisting. After months of tedious drilling, the crew hit rock at the thirty-foot level. Boring through this hard surface was even more difficult, and the crew's progress slowed down to only three feet a day. By this time, the company's investors were growing more and more disgruntled at the lack of finding oil and named the operation "Drake's Folly."

The crew was not to be deterred, however. On August 27, 1859, at sixty-nine and a half feet, a dark, green liquid spewed a few feet in the air from the well. By the second day, the well was producing eight to ten barrels of oil a day and Drake's drillers made history as the first to drill a well for the express purpose of finding oil.

History books, while covering the significance of the Pennsylvania discovery, rarely mention a second related

historical event. On October 7, 1859, the well was
destroyed when gas from the well ignited. New equipment
was moved in shortly thereafter, and the well was revived.
Thus, Drake's Folly is also recorded in petroleum history
as the first oil well fire on record. But more importantly,
the Pennsylvania Rock Oil Company had proved the
feasibility of boring through the earth's crust for a substance that was to change the world forever. The oil
"boom" was on.

SPINDLETOP

Just like the gold rush to California only ten years earlier,
oil towns sprang up literally overnight as prospectors
rushed to Pennsylvania to seek their fortunes. As the
amount of oil drilled increased, the problem of storing and
transporting the oil became a problem. But for barrel
makers, known in those days as coopers, the new industry
provided a new outlet for their wooden casks. Barrels filled
with the crude oil were often tied to rafts and floated down
Oil Creek in an effort to get them to market.

In 1861, the world's first petroleum refinery, near Oil
Creek, went on line producing primarily an illuminating
kerosene that was virtually odorless and smokeless. Railroads using flatcars transported the petroleum to refineries,
and as the market grew, so did spur lines into the oil
producing region. In 1865, a railroad tank car was specifically designed for carrying crude oil.

That same year, the first oil pipeline was laid from
Pithole City to the Oil Creek Railroad. It was two inches

in diameter and 32,000 feet long. In 1879, the first major pipeline was completed, stretching 110 miles across the Allegheny Mountains to Williamsport, Pennsylvania. By 1900, drillers were eyeing the possibility that petroleum could be found in salt domes along the Gulf of Mexico's coastline. And on January 10, 1901, in Gladys City, Texas, along the Spindletop ridge near Beaumont, a well drilled by Anthony F. Lucas struck oil, producing 100,000 barrels a day.

A second oil boom was on and scenes from the Pennsylvania oil rush nearly fifty years earlier were replayed in the hundreds of small towns along the Texas Gulf of Mexico.

PETROLEUM IN OUR EVERYDAY LIVES

By 1929, oil production in the United States had tripled from 1918 levels. It was during that year that American companies produced one billion barrels of oil for the first time in the short history of the petroleum industry.

By 1987, production of crude oil was recorded at slightly more than three billion barrels. During the past decade, the highest production recorded in one year was in 1985 when nearly 3.3 billion barrels were produced.

Drake and Lucas had paved the way for the wholesale removal of petroleum from the ground. Now, inventors began to develop products that opened up a whole new market for crude oil and its products. One of the first uses was in the internal combustion engine built in 1885–86 by Karl Benz. The concept was improved on by Gottlieb Daimler, who developed an engine that used a lighter gasoline vapor. By 1894, Rudolf Diesel had designed the engine that bears his name today, an internal combustion engine ignited by the heat of compression rather than by a spark.

However, probably the most significant invention that changed the way Americans lived came in 1893 with the development of the first American car by Charles and Frank Duryea in Springfield, Massachusetts. In 1908, Henry Ford introduced his Model T. He subsequently perfected the mass production of automobiles and America's love affair with the car had begun. This meant a completely new avenue for crude oil.

Until the development and widespread use of the automobile, gasoline was considered a waste product. However, as the number of automobiles grew, so did the number of gasoline stations across the country. The American Petroleum Institute estimates that in 1984, there were 170 million automobiles, trucks, and buses operating on the nation's 3.8 million miles of roads. That year, those vehicles consumed more than one hundred billion gallons of gasoline and sixteen billion gallons of diesel fuel.

Petroleum and its products, however, are used in many, many more ways. These include heating and cooling homes and factories, manufacturing goods, and fueling commercial and military aircraft, tractors, and railroads.

New refining processes have led to the creation of thousands of new products, including plastics, pharmaceuticals, fertilizers, paints, photographic film, detergents, synthetic rubber, and synthetic fibers. One can hardly imagine how different our lives would be today without these products.

CHAPTER 2

THE SEVEN SISTERS

The petroleum industry holds a certain mystique that other businesses can rarely match. A study of its roots reveals a parallel with the history and development of the United States and its emergence as a world leader in industry and technology.

While the fledgling industry was growing, a wealthy young man turned his few years of experience in the oil field into $1 million in cash and organized what was to be the first major oil company in the United States. No other name is as associated with those early years as John D. Rockefeller, who founded the Standard Oil Company of Ohio in 1870.

Rockefeller quickly realized the importance of petroleum and, as the industry spread to other oil producing states, so did his expansion into refineries and pipelines, as well as into overseas markets. However, public and government outcry over his monopoly of the petroleum industry led, in part, to the passage of the Sherman Antitrust Act in 1890. A lawsuit was filed against the company under the new antitrust laws, but it was not until May 1911 that the United

States Supreme Court ruled that Standard Oil must totally divest itself of its thirty-eight subsidiaries.

In fact, three of the companies eventually became part of the seven largest oil companies that collectively became known as the Seven Sisters. Their names are recognizable to anyone that drives a car and stops at a service station for gasoline and other services.

The largest of the subsidiaries was Standard Oil Company of New Jersey, which Rockefeller had formed as a holding company to escape the tentacles of the antitrust laws. Today, this company is known as Exxon, the largest oil company in the world. A second Rockefeller company was the Standard Oil Company of New York, which bought a Texas producing company called Magnolia. It later merged with another firm called Vacuum. Today, that large oil concern is called Mobil Oil.

Eleven years before Rockefeller was ordered to dissolve his vast oil empire, he had purchased a company and called it Standard Oil of California, the forerunner of today's Chevron, USA. This was the third Rockefeller company that played a major role in the U.S. petroleum industry. Another member of the early oil family was Gulf Oil Corporation, which was incorporated in New Jersey through the acquisition of the J. M. Guffey Petroleum Company and the Gulf Refining Company of Texas. These two companies were instrumental in the development of the Texas Spindletop field.

The fifth ''sister'' also had her roots in the Texas oil patch. Founded by a Pennsylvanian merchant, the Texas Company later became Texaco. The last two in the family were Shell Oil and British Petroleum, both companies founded in England.

These seven companies, which had the ability and finances to both get the petroleum out of the ground and get it to market, were to dominate the petroleum industry for decades to come. They, like others that followed and are still active in the day-to-day search for petroleum, are the major oil companies. These companies include Exxon, Shell, Texaco, Chevron, Phillips Petroleum Company, Amoco, Conoco, Sun Oil, Arco Oil and Gas, as well as many other smaller independent ones. Both of these types of companies and the personnel they employ will be covered in a later chapter in this book.

By the mid-1970s, it was clear that the destinies of the large oil concerns were beginning to be inevitably shaped by a series of uncontrollable occurrences throughout the world, particularly in Middle Eastern countries.

DEVELOPMENT OF THE MODERN-DAY INDUSTRY

Since that first oil well was *spudded,* or when drilling began, in Pennsylvania, thousands of oil field workers have dedicated their energies to improving the methods of getting the petroleum out of the ground. And out of the development of these new techniques grew a number of companies that provided services to the major oil companies. Thus, these firms that provided assistance to those drilling for oil came to be known as service companies.

Some of these same companies that broke new ground in the oil fields during the middle of this century have grown from small, family-owned firms, to corporate giants with offices around the world. Names like Halliburton, Schlumberger, Smith Tool, Baker–Hughes, and many others are heard each day during the course of drilling an oil well. Majors and independents work hand in hand with service companies to find the petroleum that is so crucial to maintaining our standards of living.

While the first wells were drilled on land, there was a growing belief that oil reservoirs could be found in the vast waters off the shores of the continental United States. The first oil well to be drilled in water was in 1896 from a pier in Summerland, California. The first well to be drilled in open waters, however, occurred in 1937. The well, drilled by a joint agreement between the Superior Oil Company and Pure Oil Company, was off the coast of Louisiana in Gulf of Mexico waters.

Since that time, the coasts of Texas, Louisiana, Mississippi, Alabama, and Florida have become some of the world's most active areas for the exploration and production of oil and natural gas. And drilling operations are a common sight from the icy waters of the North Sea to the warm waters of the Mediterranean. Even today, drilling rigs are moving further and further away from the shores, as new technology is enabling drillers to search even deeper for the petroleum.

Back on land, improved methods of getting the oil out of the ground are playing a substantial role in the United States' energy future. These new techniques and equipment are showcased at oil exhibitions held periodically throughout the world.

The search for petroleum is an exciting industry, providing unlimited, daily challenges in a business whose technology is constantly developing and changing.

TAKEOVERS AND MERGERS

While the Seven Sisters provided a base for the burgeoning oil industry in the United States, a variety of recent occurrences throughout the world has changed the face of that business even more. Falling oil prices on the world market, the ongoing Iran–Iraq war, and an oversupply, coupled with a lessened demand for petroleum, have caused the industry to reevaluate the way it does business.

Since 1981, a number of major oil companies have been bought out, or merged, with larger ones, setting a precedent for corporate takeovers in the industrial United States. The term ''corporate raider'' could be applied to the takeover attempts by Mesa Petroleum Company owner T. Boone Pickens, who forced a number of companies to merge for protection.

Oil company acquisitions rank at the top of the list of the biggest mergers in the history of the United States. To date, the largest corporate buyout—both oil and nonoil—was the 1984 purchase of Gulf Oil by Standard Oil of California (now Chevron USA) for $13.4 billion. Later that year, Texaco purchased Getty Oil for $10.1 billion. The company ended up in an historic courtroom battle with Pennzoil, who claimed it had made an earlier bid for the company. The case was eventually settled for billions of dollars. Earlier buyouts included Conoco by Dupont; Marathon Oil by U.S. Steel (now called USX); and Superior Oil by Mobil.

In 1988, a number of large, diversified companies decided to get out of the energy business. Tenneco, Inc., for example, announced in May of that year it was putting its oil and gas operations up for sale. The company, which sold its energy units to a number of oil companies, continues to operate in the areas of real estate, automotive, insurance, and farm equipment.

Some companies, however, were forced into takeovers at a time when drilling activity was low. This meant a change in personnel which included early retirements, layoffs, and an exodus by some workers into more stable industries. Naturally this affected the way companies conducted their business. Much of corporate America had to simultaneously become lean and mean to operate more profitably.

Currently, however, at a time when drilling activity is actually increasing, both major oil and service companies are scrambling to find enough experienced personnel to continue the search for petroleum. Americans are beginning to realize the importance of a strong domestic energy industry. And, while activity may be far below the levels of the boom years of the late 1970s and early 1980s, there will always be a need for experienced oil workers. Therefore, for a person undecided on a career, the energy industry is expected to provide a multitude of jobs into the 1990s and beyond.

CHAPTER 3

LOOKING FOR PETROLEUM

EXPLORATION

The beginning of the search for petroleum, whether on land or in offshore waters, always begins with the geologist. These earth detectives are sometimes called petroleum or exploration geologists and are trained to study, map, and interpret the many formations located beneath the earth's surface.

In the early days, geologists usually guessed at where significant reservoirs of petroleum lay beneath the earth. Natural oil seeps were usually a good indication that much more of the petroleum lay beneath the surface. Through a lot of luck, and probably more good hunches than science, these geologists were able to find petroleum in those early years. They accomplished this by studying exposed creek beds, railroad rights-of-way cuts, and canyons for clues of what might lay beneath the surface.

Fortunately, however, as the methods of drilling for oil improved, so did the methods of finding it become more scientific. Today's geologists, while still performing basically the same function, have at their disposal a wide range of tools and instruments used to accurately study and map underground rock formations.

Geologists use many sources of information to interpret their findings, including analyzing core samples, computers, seismic data, paleontology, and geochemistry. But no matter how advanced the method and equipment, this earth detective must still learn to accurately interpret data on certain types of formations that have been compiled and correlated. And, because of the value of the information he or she must keep any findings confidential until the employing company can secure the lease to drill for petroleum.

After a geologist completes the basic research on a potential oil field, a petroleum, or exploration, geophysicist is consulted to get a more detailed picture of subsurface formations.

While today's geophysicists have at their disposal a wide range of sophisticated tools, many are derived from the early days of the oil field. The first instruments used by early geophysicists included the surface magnetometer, refraction seismograph, and torsion balance for gravimetric surveying. Later tools were the reflection seismographs, gravimeters, and airborne magnetometers.

The most common method used in today's oil patch is seismology, the study of the earth's tremors. This procedure involves making sound waves, called seismic waves, on the surface of the earth. The waves may be made by controlled explosives; vibrators; weight dropping, where heavy weights are dropped to create the waves; or compressed gases, which produce bursts of energy using compressed air or propane. Once the waves penetrate the earth's crust, they reflect from subsurface rock layers back to the surface, where they are recorded. Instruments such as a seismometer pick up the signals and record the waves on magnetic tape and on sensitized paper.

Newer equipment includes the seismometer group recorder, which records the information on tape and helps eliminate noise and distortions that earlier methods could not possibly overcome. Most of the seismic data are also recorded for use on computers, which is helpful for constructing various types of maps and cross sections. Radar is also used to examine potential oil-bearing areas where the land is covered by forests or clouds.

Another more popular method, however, is the stratigraphic test well. This means that an area that may hold promise for oil explorers is drilled for a core sample. This bore hole sample is studied by geologists and paleontologists for traces of oil and gas and for fossils that might indicate the ages of the various rock strata.

While the methods used in searching for petroleum have vastly improved, oil companies continue to develop new and better techniques, including the increasing use of satellite imagery, and, of course, computers.

OPPORTUNITIES IN PETROLEUM EXPLORATION

There are many job opportunities within the exploration field in both the private and government sectors. Petroleum companies are primarily interested in persons whose backgrounds are in sedimentation, paleontology, stratigraphy, geophysics, structural geology, or a related discipline.

Although the fields of geology and geophysics are closely related, there are differences. Geologists study the composition, structure, and history of the earth's crust. They try to find out how rocks were formed and what has happened to them since their formation. Geophysicists use the principles of physics and mathematics to study the earth's internal composition, surface, and atmosphere and its magnetic, electrical, and gravitational forces.

Geologists and geophysicists usually specialize. Geological oceanographers study the ocean bottom, collecting information using remote sensing devices aboard surface ships or underwater research craft. Physical oceanographers study the physical aspects of oceans, such as currents and their interaction with the atmosphere. Geochemical oceanographers study the chemical composition, dissolved elements, and nutrients of oceans. Hydrologists study the distribution, circulation, and physical properties of underground and surface waters. They may study the form and intensity of precipitation, its rate of infiltration into the soil, and its return to the ocean and atmosphere.

Mineralogists analyze and classify minerals and precious stones according to their composition and structure. Paleontologists study fossils found in geological formations to trace the evolution of plant and animal life and the geologic history of the earth. Seismologists interpret data from seismographs, which measure small movements of the earth, and other instruments to locate earthquakes and earthquake faults. Stratigraphers study the distribution and arrangement of sedimentary rock layers by examining their fossil and mineral contents.

Working Conditions

Most geologists and geophysicists divide their time between fieldwork and office or laboratory work, according to the 1988–89 U.S. Department of Labor's *Occupational Outlook Handbook.* While in the field, geologists often travel to remote sites by helicopter or jeep and cover large areas by foot. Exploration geologists and geophysicists often work overseas or in remote areas, and geological and physical oceanographers may spend considerable time at sea.

Employment

According to the 1988–89 *Occupational Outlook Handbook,* geologists and geophysicists held almost 44,000 jobs in 1986. About half of them worked for oil and gas companies or oil and gas field service firms, many of which explore for oil and gas. Many others worked for business service and consulting firms, which often provide services to oil and gas companies.

About one geologist in six was self-employed; most of the self-employed were consultants to industry or government. And in addition, about 8,500 persons held geology, geophysics, and oceanography faculty positions in colleges and universities.

The federal government employed almost 6,600 geologists, geophysicists, oceanographers, and hydrologists in 1986. About three-fifths worked for the Department of the Interior in the U.S. Geological Survey, the Bureau of Mines, and the Bureau of Reclamation.

Other federal agencies that employ geologists and geophysicists include the Departments of Defense, Agriculture, and Commerce. Some worked for state agencies, such as state geological surveys and state departments of conservation. Geologists and geophysicists also worked for nonprofit research institutions and museums. Some were employed by American firms overseas for varying periods of time.

Training, Other Qualifications, and Advancement

Because so many millions of dollars are at stake, a petroleum company must have access to the best geologic and geophysical data. More importantly, they must have qualified geologists and geophysicists who are able to perform their primary function, which is to properly interpret that data.

A bachelor's degree in geology or geophysics is adequate for entry into some lower-level geology jobs. But better jobs with good advancement potential usually require at least a master's degree in geology or geophysics. Persons with strong backgrounds in physics, mathematics, or computer science also may qualify for some geophysics jobs. A Ph.D. degree is essential for most research positions.

Over 500 colleges and universities offer a bachelor's degree in geology or geophysics. Other programs offering training for beginning geophysicists include geophysical technology, geophysical engineering, geophysical prospecting, engineering geology, petroleum geology, and geodesy. More than 270 universities award advanced degrees in geology or geophysics.

Geologists and geophysicists need to be able to work as part of a team. They should be curious, analytical, and able to communicate effectively. Those involved in fieldwork must have physical stamina.

Geologists and geophysicists usually begin their careers in field exploration or as research assistants in laboratories. They are given more difficult assignments as they gain experience. Eventually, they may be promoted to project leader, program manager, or other management and research positions.

Job Outlook

Data compiled by the U.S. Department of Labor indicate that employment of geologists and geophysicists is expected to increase five to thirteen percent through the year 2000. Most jobs for geologists and geophysicists are in or related to the petroleum industry and, in particular, the exploration for oil and gas.

This industry is subject to cyclical fluctuations, and exploration activities have been greatly reduced because of the drop in the price of oil. Consequently, recent employment prospects for many geologists and geophysicists have been poor. Labor Department statistics indicate that employment opportunities will remain poor until oil prices increase enough to make more exploration worthwhile. Since new sources of oil and gas eventually must be found, exploration activities should increase in the future. When this occurs, there will probably be excellent employment opportunities because many experienced geologists and geophysicists have left the industry. At the same time, the number of degrees granted in geology probably will be greatly reduced until opportunities improve.

It is difficult to predict when oil prices and exploration will increase. Some analysts expect the price of oil to increase enough by the early 1990s to encourage more exploration.

Earnings

Surveys by the College Placement Council indicate that graduates with bachelor's degrees in physical and earth sciences received an average starting offer of $19,200 a year in 1986. Experienced earth scientists with a master's and/or doctorate may earn around $40,000 annually.

In the federal government, in early 1987, geologists and geophysicists having a bachelor's degree could begin at $14,822 or $18,358, depending on their college records. Those having a master's degree could start at $18,358 or $22,458 a year; those with a Ph.D. degree, at $27,172 or $32,567. In 1986, the average salary for geologists in the federal government was about $37,500 a year, and for geophysicists about $40,900 a year.

Related Occupations

Many geologists and geophysicists work in the petroleum and natural gas industry. This industry also employs many other workers in the scientific and technical aspects of petroleum and natural gas extraction. These workers include drafters, engineering technicians, science technicians, petroleum engineers, and surveyors. Also some physicists, chemists, and meteorologists, as well as mathematicians, computer scientists, and cartographers, do related work.

Companies that provide equipment and support services to those looking for petroleum also offer an abundance of opportunities. Needed are personnel to read the core samples, operate the seismic equipment, and punch the computers. Most of these on-the-job training opportunities will be covered in chapter 8.

LEASING LANDS

Before any drilling, exploration, or production can take place on any land suspected of being a good oil prospect, the mineral rights must be secured from the property owner. This often complex but critical job is known as leasing, and is the duty of a landman. This negotiator of lands must learn not only oil field terminology, but also federal, state, and local laws in order to execute a legal, binding contract. He or she must also be something of a social scientist who can convince a reluctant landowner of the benefits of having his or her property explored and, with luck, produced.

Landmen, who are specifically concerned with the legal rights to any hydrocarbons found beneath the subsoil, must deal primarily with two kinds of properties: public (owned by the government) and private (owned by an individual or corporation). A third type of property are those lands owned by the American Indian.

However, no matter who owns the property, a landman must be knowledgeable in the particular laws of the land, as well as adept in negotiating a contract satisfactory to all parties involved.

Leasing Private Lands

While a private landowner has the right to search for and remove any minerals from his or her own property, most, obviously, have neither the finances nor the expertise to do so. Therefore, a landowner might decide to lease or sell all or part of his or her land rights to someone else. The primary function of a landman, then, is to search the title of the property, review the company's anticipated operations with the owners, negotiate signatures to the mineral lease, and have the contract recorded with the correct legal entity.

Most of the time the landman will begin negotiations with property owners after the geological work has been done. However, in some instances, the leasing process begins prior to beginning the geophysical work. This usually happens when property is located in an extremely promising area where successful, producing wells have been drilled.

In either case, the landman stays in constant contract and continues to work closely with the geologist and/or geophysicist.

A landman can offer a private landowner three options on his or her land: a lease interest, mineral interest, or royalty interest.

In an oil and gas lease, a written agreement is made between the landowner and the oil or gas company. This contract, usually written for a specific number of years, gives the company exclusive rights to enter the land, prospect for petroleum, and drill and remove any petroleum found there. In return, the landowner is given an initial bonus for agreeing to lease the land. Additionally, he or she gets a yearly rental fee based on how much acreage was leased. And, if the land does produce oil and gas, the landowner shares in the production. This is called a royalty payment and is usually 12 1/2 percent—or one-eighth—of the value of oil and gas produced on the lease. All the while, the property owner can continue to use the surface of the land, as long as it does not interfere with the company's operations.

In a mineral interest option, a property owner can either sell the rights to some or all of his or her minerals, or sell the land and keep some or all of the mineral interests. This is where the job of a landman becomes more specialized and complicated, as the rights of either property owner may vary from state to state.

The last option a landman can offer a property owner is called royalty interest. This means a landowner agrees to transfer only the rights to the proceeds from the minerals.

As you can see, the job of a landman is not only extremely important, but very involved as well. There are many variations of lease options to negotiate and federal, state, and local laws to follow. A landman must be well versed in all of them.

Leasing Federal Lands

At the beginning of fiscal year 1985, the federal government owned more than fifty percent of the total onshore and offshore areas of the United States. Onshore, the government owned 725 million—thirty-two percent—of this country's 2.3 billion onshore acres and controlled the mineral rights to another 66 million acres. Offshore, the federal government owned the subsurface rights from the Outer Continental Shelf (OCS), a band three nautical miles from the coastline, to a 2,500-meter water depth, or 966 million acres. Coastal states owned the remaining 33 million acres.

Lands owned by the federal government or state governments are usually acquired through a competitive bidding process involving numerous and specific rules and regulations. Federal onshore leasing is conducted through both competitive and noncompetitive methods. Competitive leases of up to 640 acres and five years are granted on acreage located on a known geological structure. If the land is not located on a known geological structure, terms for noncompetitive leases include granting a company acreage up to 2,560 acres and ten years.

The most visible types of federal land sale are the offshore lease sales that are periodically held for acreage in the OCS. These lands are leased through the U.S. Department of the Interior's Mineral Management Service. Announcement of the impending sale is published at least one month in advance in the government's *Federal Register*.

Prior to the date of the sale, companies submit sealed bids for the tracts, which normally have a five-year term.

If the company whose bid was accepted does not conduct initial exploration operations as specified in the lease, the acreage reverts to the federal government, and the company loses the money paid to purchase the tracts.

Petroleum companies that do not own acreage in a certain offshore tract may try to participate in a *farm-in*. This means that the company would share an existing lease held by another company and be responsible for some or all of the cost of drilling a well. In return, the company would get a specified interest in the acreage. The opposite of a farm-in is a *farm-out*. In this type of leasing arrangement, a company already holding acreage would take on and pay drilling partners to explore certain portions of its holdings.

Whatever type of leasing arrangement is used, companies bid only on tracts where extensive geological tests have been conducted and analyzed and the prospects for a large find is possible. Even with the downturn in the industry for the past few years, the cost of drilling one offshore well continues to be in the millions of dollars. And, as the petroleum industry continues its quest for new oil frontiers, new technology is being applied to wells that are being drilled deeper, further from land, and in even more remote locations far from the comforts of civilization.

OPPORTUNITIES FOR LANDMEN/LEASING AGENTS

The term *landman* is most likely a throwback to the early days in the oil field when few women, if any, were employed in the fledgling business. However, in today's oil patch (and in this book), the term is applicable to both sexes.

Getting property owners to agree to have their land drilled and produced was, in fact, one of the first career opportunities in the petroleum industry that became available to women. Today, it continues to be a source of employment for thousands of women who have learned the art of negotiating and closing a deal.

Landmen who work for major oil companies are sometimes called land agents. But whatever their title, their ultimate responsibility remains the same: to secure the rights to explore, drill, and produce on tracts of land or the ocean bottom that earth scientists believe to be promising.

Training, Other Qualifications, and Advancement

Some colleges, primarily in the energy producing states of Texas, Louisiana, and Oklahoma, offer degrees in petroleum land management. Most companies prefer an employee with a law degree since some land agents are also used to negotiate exploration or development contracts with other producers. Or, they arrange farm-in or farm-out agreements.

A natural step from a land agent's job is a career in the company's legal affairs or general management departments.

Job Outlook

According to 1987 survey figures from the Lafayette area Job Service Office of the Louisiana Department of Labor, a landman could have earned about $35,000 annually, while a landman's assistant could have earned about half that amount.

Some landmen and geologists may be interested in the challenge of owning their own company, where they can do reasonably well. However, it would probably be a good idea to first get hands-on experience by working for several years with a company. This would not only provide on-the-job training, but an opportunity to build up the nest egg necessary for opening your own office.

Before setting out on your own, however, it would be wise to check with other professionals in the field in your particular part of the country. Because of the cyclical nature of the petroleum industry, it may be more prudent at a certain point in your career to remain in the security of a company, rather than taking the risk of going out on your own.

Trends in your particular area of specialization may also be checked out with the associations, schools, and organizations listed in the appendixes in the back of the book.

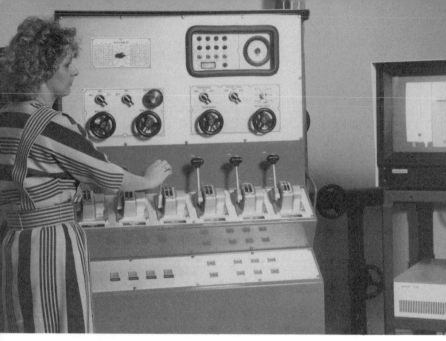

Top: Petroleum workers can learn how an oil or natural gas well is drilled by using a drilling simulator. (photo by Dennis Sullivan, Lafayette, Louisiana) *Bottom:* This crew handles drilling equipment in the Gulf of Mexico (American Petroleum Institute photo)

CHAPTER 4

DRILLING A WELL

Although the methods of determining where the most promising pockets of hydrocarbons are located have vastly improved since the days of Colonel Drake, the only sure way to find out is to actually drill a well.

Drilling an exploratory oil or gas well is costly and much planning goes into a drilling program before the first hole is ever punched. There are three types of exploratory wells: those drilled to find the limits of an oil or gas bearing formation; those drilled in search of a new productive formation in an area that already contains commercial wells; and new field *wildcats,* which are those exploratory wells drilled where neither oil nor natural gas have been found before.

The cost of drilling one wildcat well can range from half a million dollars to more than $20 million, depending, of course, on where the well is drilled and how deep it will be drilled. A company can absorb all of the costs of drilling a wildcat well, or if a company is incorporated, it can be financed from the sale of company stock. An operator may strike a deal with other companies, even competitors, to share the risks or rewards if the well is successful.

Most of the early wells were quite shallow. Remember the Drake well, which was sixty-nine and a half feet deep and took less than fifteen days to drill? Today's onshore wells are drilled to an average depth of nearly five thousand feet. Many more go beyond fifteen thousand feet. One of the deepest oil wells ever drilled in the United States was in Oklahoma, a dry hole drilled to 31,441 feet. A well being drilled in the Soviet Union is said to be approaching the forty-thousand-foot mark!

Obviously, the deeper a well is drilled, the longer it takes to drill and the higher the costs. As the price of oil increased in the late 1970s and the early 1980s, so did the cost of drilling. In 1975, for instance, the average cost of drilling one oil or gas well was $177,793. By 1986, the average cost of drilling an oil or gas well, including a dry hole, had jumped to about $400,000 per well.

According to a survey compiled by the American Petroleum Institute, the Independent Petroleum Association of America, and the Mid-Continent Oil and Gas Association, companies spent $9.2 billion in 1987 to drill and equip onshore oil and gas wells and dry holes. That's a thirty-one-percent drop from the $13.6 billion spent in 1986. The average per foot cost of drilling a well also dropped in 1987. Dry hole costs fell by one-third, gas well costs fell twenty-five percent, and oil well costs were down twenty-five percent, according to the survey. In 1987, the total cost of drilling offshore wells and dry holes amounted to about $2 billion. That's a decrease of nearly forty-three percent from 1986. And by the end of 1987, officials counted 923 active rotary rigs at work in the United States. At the end of 1988, there were still less than one thousand rigs operating. The highest number of rigs to ever work at once in the United States was 4,500, which was noted in December 1981.

Although most wells are drilled vertically, some operators use the directional drilling technique. This method of drilling allows several wells to be drilled from one location. This is obviously an advantage, especially offshore, where the cost of one platform may run into the millions of dollars.

Most of the explorers who drill new oil and natural gas wells in the United States are the smaller drillers of the industry who are called independents. These wildcatters drill over eighty-five percent of all wells and ninety percent of the exploratory wells in the United States. An independent can be an individual, partnership, or public corporation.

At the peak of the oil boom there were about fifteen thousand independents. Today, there are about twelve thousand of these wildcatters who spend an average of $200,000 per well to drill what is historically successful only one out of five times. The other four wells, on the average, are the dry holes. Only one in sixty wells is considered a significant discovery.

A typical independent does not own refineries or gasoline service stations. The independent runs his or her own business, calling the shots on whether to drill. While the name of independent drillers may not be as familiar as that of the larger companies known as majors, their role in the exploration and production of new oil and gas reserves is of great significance.

Independents:

- drill ninety percent of all exploratory wells in untested areas, seeking new reserves of oil and natural gas;
- produce about a third of the total output of crude oil and natural gas in the United States;
- operate most of the 450,000 small oil wells in twenty-eight states that produce an average of less than three barrels a day. These wells, known as stripper wells, together produce 1.2 million barrels of oil a day, fifteen percent of U.S. crude oil output, from fields containing twenty-one percent of the U.S. total proved oil reserves in the lower forty-eight states; and
- make eighty percent of the significant U.S. oil and natural gas discoveries, according to the American Association of Petroleum Geologists.

Independents sell their crude oil to refineries, which in turn market the refined products through their own outlets or through those of independent marketers. Natural gas reserves discovered and produced by independents are sold at the wellhead to natural gas pipelines, which in turn sell it to local gas utilities for distribution to consumers.

Major petroleum companies can rely on other sources of revenue from their companies when oil and natural gas prices fall below a profitable level. Independent producers, however, live and die by the price they receive at the wellhead.

DRILLING ONSHORE

Once an independent or major company decides to drill a well on land, it must first select a well site that is easily accessible, as level as possible, and located so the drilling process will have as little effect on the surrounding environment as possible. The company, known as the operator, then invites drilling contractors, those companies that own and operate rigs, to bid on the job. Once the job is awarded and all costs negotiated, a contract is signed between the operator and drilling contractor.

Personnel who will be involved with the drilling are brought in, and the site is cleared and leveled in preparation for placement of the drilling rig. In some cases, turnarounds and access roads built of planks called board runs are built, while support facilities are brought in for the crew. If the area is not easily accessible, heavy lift helicopters are used to bring in the equipment. All of this obviously means the property will be altered. However, laws and regulations require that the change be minimal and repairable. The land must be restored to its natural state after drilling.

Large amounts of water are needed during land drilling operations. If a stream, river, or other suitable source of water is nearby, pumps and a waterline are installed to bring water from that source to the drill site. If this source of water is not readily available, a water well is drilled.

Because of regulations and an increased awareness of the environment by oil companies, there has been more care taken when disposing of materials considered potentially hazardous to the area. Part of the preparation includes digging an earthen pit, which is then lined with plastic. This pit is used to collect used or unneeded nontoxic drilling materials. After the well is drilled, the pit is covered and leveled.

If the well is located in an ecologically sensitive area, little, if any, waste is dumped at the site. Trucks are used to haul off the material to locations that have been approved for disposal.

DRILLING OFFSHORE

While the concept of drilling an offshore well is basically the same as drilling one on land, they are very different situations that must be dealt with by companies expert in drilling in several hundred feet of water.

After the drilling contractor is awarded the contract, preparations of the offshore site are made. Buoys are set to mark the spot where the operators want to drill. Some offshore rigs sit atop platforms. Therefore, these rigs must be moved from land and loaded onto large barges for the slow trip to the site. In order to be moved, the rig is usually divided into sections, called packages, which are hoisted individually by a gigantic crane onto the barge. The rig is then reassembled on the platform. This requires precise planning, since all supplies needed to rig up will have to be brought along on the trip.

TYPES OF RIGS

While all onshore wells are basically drilled the same, the type of offshore well to be drilled determines the kind of rig that will be used. Depending on the depth of the water, climatic conditions of the area, and the costs involved, any one of five kinds of mobile offshore drilling units (MODU) can be used to drill exploratory wells.

Submersible Rig

One of the oldest types of MODU used for exploration is the submersible rig. Its design probably came from the first offshore rigs, which were sunken barges that were secured in place by wooden pilings. This MODU has an upper level that houses the crew and working area, and a lower level, the hull, which sits on bottom while drilling. The hull is pumped out and floated to move the rig to a new location.

Submersibles drill in shallow water of thirty to forty feet, although newer units are able to drill to one hundred feet. Once exploration is completed, it is moved to a new site.

Semisubmersible Rig

One of the most common types of rigs used is the semisubmersible. While some models may sit on the bottom in shallow water, others are more frequently used in a floating, or partially submerged position, and held in position with anchors. Semis are used primarily in deep, rough waters up to two thousand feet. Once a well is completed, the unit is deballasted and moved on to the next location. Some newer models are self-propelled.

Jackup Rigs

Jackup rigs are used to drill in water depths of up to about four hundred feet. It is one of the most widely used, bottom-supported MODU. The rig is built to float to its location while being towed. Once it has arrived at its destination, the rig's huge legs are lowered to the seafloor, and its hull is jacked up on the same legs to raise the drill deck above the water's surface.

Drill Barges

Another type of drilling unit used in shallow and inland waters is the inland drilling barge. This unit consists of a drilling rig mounted on a barge and towed by a tug. Any movement of the barge must be made under the power of another vessel. Barges also house equipment, supplies, and crew quarters, although other services associated with drilling a well are usually furnished from other barges or vessels.

Drill Ships

This type of rig is used primarily in deep water and can carry larger cargoes of drilling and maintenance supplies, making it more suitable for remote locations. Once a well is completed, this MODU, because it is self-propelled, is able to move under its own power to a new location.

After a MODU has confirmed the existence of hydrocarbons in a particular location, developmental drilling may be planned. These locations are sometimes drilled from a rig sitting atop a fixed offshore platform. Platforms are constructed in onshore fabrication yards. These platforms must be designed and constructed to withstand the extreme forces of nature, such as hurricanes in the Gulf of Mexico, icy winds in the Arctic Sea, or earthquakes in the Pacific. Once constructed, they are loaded on barges and towed out to a location. Once at a site, they are lowered to the bottom and securely anchored to the seafloor with pilings, which are driven deep into the soil. Huge cranes lift the rig packages from the barge to the platform, where several wells can be drilled by sliding a movable derrick to drilling slots built into the platform.

Since the first platform was installed in the twenty-foot waters of the Gulf of Mexico in 1947, thousands of fixed platforms have been placed throughout the world's prolific offshore areas. About two-thirds of all fixed platforms are located in the Gulf of Mexico, where some have been placed in well over one thousand feet of water.

COMPONENTS OF A DRILLING RIG

It is up to the contractor drilling the well to provide the proper equipment and personnel to operate a rig as agreed upon in the contract. The main function of a rotary drilling rig, the most common type of rig used, is to drill a well, or ''make hole,'' as it is called in the industry. All drilling rigs have four main components: power, hoisting, rotating, and circulation systems.

Power System

Most all rigs use a diesel-fueled internal combustion engine as their main source of power. A rig's engine is not unlike a car's engine, only much larger and much more powerful. Most rigs are powered by at least two engines that provide from five hundred to three thousand horsepower.

Hoisting System

This system is composed of five parts: The *hoist* is used to run the drill pipe in and out of the hole to change drill bits or run tools. This drill string (drill pipe and drill collars) must be lifted out of the well, disconnected and racked back, the bit changed, and run back in the hole. The *derrick* is designed to hold 250,000 to 1.5 million pounds of pipe and withstand winds up to 130 miles per hour with the racks full of pipe. The drilling line is wound around a revolving drum called the *drawworks,* which houses the main brake used to hold and stop the many pounds of pipe being raised and lowered into the well.

The *traveling block* and *crown block* are each a series of large pulleys through which the drill line travels. This system of multiple lines increases the hoisting ability of the drawworks.

Rotating System

This component is comprised of all the parts that extend from the swivel to the drill bit. All of the parts move, providing motion for the drill bit.

The *swivel* carries the weight of the drill string, permits it to rotate, and provides an opening for the circulating fluid. The *kelly* is a hollow, six-sided tube that helps keep the drill string moving vertically as it is lowered during drilling operations. Providing the torque to turn the drill stem and space for slips to hold the pipe is the *rotary table*. Mud is pumped through thirty-foot sections of *drill pipe* and *drill collars* to which is attached the *drill bit,* which is used to break up the formation. There are a variety of drill bits on the market and what type is used depends upon the formation being drilled.

Circulation System

Drilling fluid, called mud in the petroleum business, is usually a mixture of water, clay, weighting material, and a few chemicals. It is used to raise cuttings made by the bit and lift them to the surface for disposal. But more important, perhaps, is that it helps keep underground pressures in check. It is part of several safety measures used to protect workers from a blowout. This safety measure, as well as others, will be covered later in this chapter. Fluid circulating equipment and mud are major costs associated with drilling a well. Other equipment is needed to maintain the consistency of the drilling mud, as well as equipment for removing fine solids.

Drilling operations begin by attaching a bit to the drill string. This drill string passes through a turntable on the derrick floor, and as the pipe is lowered and rotated into the earth, the bit rotates and bores deeper and deeper into the ground. As the hole is drilled, new pipe is added. When drill bits become dull or break, the entire drill string must be pulled out, the bit changed, and the drill string run back in the hole. This process of removing, reconnecting, and then resuming drilling happens over and over again when a deep well is being drilled.

SAFETY AT THE WELL SITE

While developing technology has kept the number of blowouts to a minimum, there is still a large amount of danger associated with drilling a well. New employees are instructed in the proper safety rules and regulations of working on an offshore drilling rig. Specific duties are also assigned in case the rig has to be suddenly evacuated. Periodic safety drills are also held and all personnel must be familiar with escape routes.

Most offshore rigs are equipped with a helipad for the use of helicopters to ferry crews on and off the rigs. It can also be used for evacuation purposes. Most offshore rigs are also equipped with survival capsules and employees are instructed in how to properly board and launch them. Other offshore survival techniques are taught periodically to workers, such as how to escape from a helicopter that has overturned in the water.

So concerned is the petroleum industry about offshore safety that many major oil companies have contributed to the establishment of the United States' only Marine Survival Training Center. The facility was built in conjunction with the University of Southwestern Louisiana and is located in Lafayette, Louisiana. It was built in response to

Coast Guard regulations that went into effect in early 1989 and require that several individuals on every offshore installation be trained in launching and navigating survival capsules.

The center is the nation's first large-scale training facility for training personnel to operate the capsules. Instruction includes launching, piloting, and retrieving survival capsules, as well as using them in fire and hydrogen sulfide emergencies.

Many petroleum workers have been injured and killed by blowouts, which occur when well pressure is not controlled. However, improved drilling techniques, as well as new equipment, help monitor and control well-bore pressure. On both land and offshore rigs, a blowout preventer is installed under the rig floor. Valves to close the well can be controlled from a remote panel, should it become necessary. If a blowout is serious, rig workers are evacuated and experienced wild-well fighters, such as Red Adair, are called in to shut in the well.

DRILLING OPPORTUNITIES

Because of the slowdown in the oil patch over the past several years, there has gradually been a shortage of drilling personnel. In recent years, many experienced professional and technical workers have given up on this cyclical industry and have gone on to other lines of work. Therefore, should the industry begin its predicted, gradual upturn during the early 1990s, the outlook for careers in drilling should increase proportionately.

A large number of people are involved in directly drilling a well. A large drilling contractor may own fifty rigs, while smaller ones may own only a handful. Larger contractors may have rigs operating all over the world. Usually, area or regional offices are set up in strategic locations where rigs may typically operate.

Jobs in Drilling

A regional manager, drilling superintendent, and drilling/production engineer are usually responsible for several rigs. They work in the regional office and keep in daily contact with the rig by telephone or radio, as well as by daily reports.

Drilling/production engineers direct drilling and production operations. They are also responsible for overall planning, selecting equipment, drilling methods, and recovery methods. They also determine the most desirable rate of oil flow from the well.

Working on the rig is a toolpusher, who is actually in charge of the work site. Drilling operations are conducted around the clock in shifts called *tours* (pronounced "towers"). Offshore personnel generally work twelve-hour shifts, seven days on and seven days off. Onshore, the shifts are eight hours, and they generally work five days on, with two days off.

The head of each tour is a driller, who heads a crew comprised of a derrickman and three roughnecks. The driller's job is one of the most important on a rig and requires several years of experience. He or she is usually promoted to the driller's position after working several years each as a derrickman, floorhand, and motorman.

Drillers work closely with the toolpusher and company representative while carrying out the drilling program. A driller must be able to make quick decisions during drilling operations, and, if a blowout or accident occurs, must calmly handle the emergency and administer first aid. A driller is also the crew's morale booster, as well as the person who sometimes assists with personnel recruitment.

A derrickman may be considered something of a daredevil. Attached with a safety belt about ninety feet above the drill floor, a derrickman racks the drill string as it is pulled from the well. He or she also handles the drill string when it goes back into the hole. In most cases, he or she is also responsible for the mud system and mud pumps. The danger of this job is compounded by high winds and, if an accident such as a blowout occurs, the derrickman may get caught up in the derrick.

Floormen, better known as roughnecks, perform most of the physical work during drilling operations. They are responsible for screwing and unscrewing the joints of pipe as they come out of the hole. They must also wash down and rack back the pipe as it comes out, making it one of the dirtiest and greasiest jobs on a rig. Since they work outside and are unprotected from the elements, roughnecks must enjoy the outdoors and be able to adapt to the rigors of any climate.

A number of drilling personnel began their careers as a roustabout, a person who assists with the drilling operations and performs general maintenance on a drilling rig.

Offshore rig workers live and work on offshore rigs during their shift, usually called "seven and seven." That means a rig worker will work seven straight days and then will be off seven days. Some overseas jobs call for "fourteen and fourteen" or "twenty-eight and twenty-eight" shifts. Land rig personnel may live in trailers moved in near the well site or at other close quarters provided by the company. But whatever the work schedule, the rig workers, supplemented by personnel from service companies called in to perform certain operations, actually drill the well.

Oil field personnel work hard, but they are also well paid for their services. Key positions and approximate annual salaries include:

- drilling superintendent: $47,000
- toolpusher: $39,000
- driller: $32,000
- derrickman: $24,000
- roughneck: $22,000
- motorman: $22,000
- crane operator: $22,000
- roustabout: $20,000
- electrician: $30,000
- mechanic: $30,000
- welder: $26,000

The opportunities for working on a drilling rig depend, of course, on the amount of drilling activity planned by both major oil companies and independents. And, at this point, since exploration is contingent upon the rise in oil prices, it is difficult to predict what kind of opportunities these positions will present in the near future.

EMPLOYMENT FOR PETROLEUM ENGINEERS

Petroleum engineers are responsible for exploration and drilling for oil and gas. Many plan and supervise drilling operations. The U.S. Department of Labor's 1988–89 edition of the *Occupational Outlook Handbook* reports that petroleum engineers held almost 22,000 jobs in 1986, mostly in the petroleum industry and closely allied fields. Employers include major oil companies and hundreds of smaller and independent oil exploration, production, and service companies. Engineering consulting firms, government agencies, oil field services, and equipment suppliers also employ petroleum engineers. Others work as independent consultants.

Many petroleum engineers are employed in Texas, Oklahoma, Louisiana, and California, including offshore sites. Also, many American petroleum engineers work overseas in oil-producing countries.

Training, Other Qualifications, and Advancement

A bachelor's degree in engineering from an accredited engineering program is generally acceptable for beginning engineering jobs. College graduates with a degree in science or mathematics and experienced engineering technicians may also qualify for some engineering jobs, especially in engineering specialties in high demand. Because of the shortage of qualified petroleum engineers in recent years, some companies have hired other types of engineers and trained them specifically for the oil and gas industry. This is an advantage to workers since it allows them to shift to fields with better employment prospects, or ones that match their interests more closely.

Graduate training is essential for engineering faculty positions but is not required for the majority of entry-level engineering jobs. Many engineers obtain a master's degree, however, because it often is desirable for learning new technology or for promotion. Nearly 260 colleges and universities offer a bachelor's degree in engineering, and nearly 100 colleges offer a bachelor's degree in engineering technology. All fifty states and the District of Columbia require registration for engineers whose work may affect life, health, or property.

Beginning engineering graduates usually perform routine work under the close supervision of experienced engineers, and in larger companies, may also receive formal classroom or seminar-type training. Engineers should be able to work as part of a team and should have creativity, an analytical mind, and a capacity for detail. Additionally, engineers should be able to express themselves well, both orally and in writing.

Job Outlook

The Center for Education Statistics said 1,719 bachelor's of science degrees in petroleum engineering were granted in academic year 1984–85. During that same year, 265 petroleum engineers received their master's of science degrees, while 24 earned doctorate degrees.

Employment of petroleum engineers is expected to grow five to thirteen percent through the year 2000. With the drop in oil prices, domestic petroleum companies have sharply curtailed exploration and production activities, resulting in poor employment opportunities for recent petroleum engineering graduates. In the long run, however, it appears likely that the price of oil will increase to a level sufficient to increase exploration and production, which should improve employment prospects for petroleum engineers. Despite this expected employment growth, most job openings will result from the need to replace petroleum engineers who transfer to other occupations or leave the labor force.

Petroleum engineers may find career opportunities in oil and gas exploration and production companies; oil and gas service companies; financial institutions; general engineering practice; education and training; and geothermal, chemical, and agriculture industries. Other areas of employment include government; the mining, pipeline, and computer industries; research; gas processing; consulting; and law.

Earnings

Petroleum engineers are the highest paid of all the engineering professions. In 1986, the average starting salary for a petroleum engineer with a bachelor's of science degree was $33,000. That's substantially higher than the $27,900 average starting salary calculated by the College Placement Council. Those with master's degrees and no experience averaged an annual salary of $33,100, while those with a Ph.D. averaged $42,200.

A geological engineer, whose job concerns applying principles of rock and soil mechanics for engineering projects, may earn nearly $35,000.

Related Occupations

Engineers apply the principles of physical science and mathematics in their work. Other workers who use scientific and mathematical principles include physical scientists, life scientists, mathematicians, engineering and science technicians, and architects.

CHAPTER 5

PRODUCING AND RECOVERING PETROLEUM

Since the cost of drilling and producing a well is so expensive, operators must decide whether it will be profitable to bring the well into production. The Society of Petroleum Engineers, in its *Oil from the Earth* booklet, says this decision involves part or all of four steps:

- evaluation of the formation
- isolation of the formation
- stimulation of the well
- installation of production equipment

It is the job of the petroleum engineer to plan and execute all of these steps.

It is the job of a production engineer to optimize hydrocarbon production from existing wells through the use of chemical or mechanical stimulation. They may also be involved in re-evaluating the method used to lift the oil and gas out of the reservoir or in sizing and selecting separation, handling, metering, and disposal equipment as part of the surface treating facility.

To evaluate the formation requires the services of a mud logger, who has been monitoring downhole conditions while the well was being drilled. Using a microscope or ultraviolet light, the mud logger studies cuttings made by the drill bit from the rock formation in the well below. From these cuttings, he or she is able to determine whether oil is present. Another tool used by the mud logger for the same purpose is a gas detection instrument.

One of the most innovative methods used today, however, is the well-logging technique. A piece of equipment called a logging tool is lowered into the well, and then slowly reeled back to the surface. As it is brought back up, the tool is able to measure the properties of the formations through which it has passed. Electric logs work by measuring and recording resistivity in the formations, while some logs use sound waves to help define the formations.

Another popular device measures the formation pressure as fluids enter the tool. Core samples are also used to determine the presence of hydrocarbons in the particular formation. Other methods use a radioactive source to define formations.

COMPLETING A WELL

If no hydrocarbons are present, or if an operator feels the cost would be prohibitive to produce the well, he or she may decide to seal off the well, using a number of methods known as plugged and abandonment. Government regulations require that both onshore and offshore well sites be cleared of all debris and wells sealed with cement plugs to prevent fluids from escaping from or entering the well. Some areas require that the land surrounding onshore wells be replanted and reworked. In some cases, there is little evidence that a well had ever been drilled.

To prevent problems for commercial vessels, some off-shore wells are plugged and cut off below the seafloor. However, in some coastal areas, abandoned offshore platforms are left in place or moved to another area to be used as artificial reefs. Fish and other sea creatures are attracted to the structures, thus providing a better than average sport-fishing environment.

If a well holds any promise at all, and the price of oil or natural gas warrants it, operators sometimes go back and produce a well that has been plugged and abandoned. Operators have a tough decision to make when deciding to produce a well, since the cost of bringing a well into production can run into many thousands of dollars. If an operator determines that a well would be economically feasible to produce, he or she begins preparations for production. This is known as completing a well.

Pipe called production casing is set and cemented in the formation to isolate the pay zone (productive area). To get the well to flow up to the surface, a technique known as perforating is used. The most common method of perforating is to fire charges through the production casing and into the formation. This provides a means for the fluid to get into the well bore so the hydrocarbons can flow into it and up to the surface. A piece of equipment built with a series of arms and valves and called a Christmas tree is attached to the wellhead. This device controls the well and the flow of oil or gas into the gathering pipelines.

Different types of wells present different completion problems, and operators have alternate methods at their disposal for completing a well.

RECOVERY METHODS

Scientists estimate that only about one-fourth of the oil in a reservoir is recovered by natural flow and pumping. So, most wells need a little help in getting the oil to flow at a slow, steady rate.

Reservoir engineers are usually called in to make decisions about major development phases for a reservoir. They must select the method most economically beneficial to the reservoir by calculating the amount of recoverable oil and gas. Then, they must determine the number of wells economically justified to recover those reserves. Many now use computers to simulate future performance for better decisions about the reservoir.

One of the most common types of primary recovery methods used is the artificial lift, including the use of a pumping unit. This piece of equipment, sometimes called a horse's head because it looks like a gigantic rocking horse, is required for most wells when natural pressure diminishes. Another artificial lift technique used is the circulation of natural gas, which forces the oil to the surface. Special pumps can also be used to push the flow of oil through the well.

Secondary recovery methods developed in recent years include water flooding and chemical flooding, as well as thinning the oil by heating to make it flow easier. The newest development in oil production is called enhanced oil recovery. There are several types of recovery methods, but all center on injecting certain chemicals or gases that mix with the oil. Thermal recovery uses steam injection to thin thick crude oil, thereby making it easier to produce. Another method is called fire flooding. This is accomplished by starting a fire in the reservoir, which is fed by injected air. According to the Society of Petroleum Engineers, forty percent of the oil produced in the United States now comes from fields where reservoir injection processes are applied. This compares to a fifteen to twenty percent recovery rate through primary recovery methods, the SPE estimates.

SERVICE AND MAINTENANCE

After production equipment is installed on a producing well, it must be periodically serviced to maintain or improve production. This is called well servicing. A more extensive type of maintenance is called workover. Using a scaled-down version of a drilling rig, a high pressure pump forces liquids through the tubing to the bottom of the well, outside the tubing, then back to the surface. This allows the well pressure to be controlled from the surface and work can then be performed on the well. There are companies that specialize entirely on well servicing or workover maintenance.

Hydrocarbons are removed from the ground in complex mixtures of water, oil, and natural gas, and must be separated and treated at the well site. Two tanks are usually used since one will be filled while the other is being gauged for its oil content and emptied into a tank truck or pipeline.

Large pipelines carry the natural gas production from the well to market, while the water is sometimes reinjected into the field to help maintain reservoir pressure, which helps in the production of more oil. Natural gas is widely used for industrial purposes, as well as by most consumers for cooking and heating. Therefore, large supplies are kept near market areas for long-range use. Oil and natural gas produced offshore must also be brought ashore by pipelines, which are required to be buried below the mud line in water depths less than two hundred feet.

Specially equipped semisubmersible barges are now able to lay lines in water depths ranging to one thousand feet, while vessels using reel-type equipment can lay lines in water depths of three thousand feet. In shallow waters with mild sea conditions, the oil is stored in tanks and transferred to barges.

How petroleum is transported from the well site to refineries and the marketplace will be detailed in chapter 6.

EMPLOYMENT OPPORTUNITIES

Professional Jobs

The many companies that perform the services listed above employ a variety of personnel for both in-house and field work. Students interested in any type of engineering careers should concentrate on math, English, physics, and chemistry. Other important courses are computer science, geology, mechanical drawing, economics, and government or social studies.

For professional jobs in the production field, a bachelor's, master's, or doctoral degree is required. The search for petroleum is becoming increasingly complex, requiring personnel with advanced education and experience. Many professionals such as engineers, geologists, or geophysicists, are now being given management opportunities. Some are involved in the day-to-day production activities, while others hold support staff engineering positions.

Persons interested in pursuing an engineering career should have good problem-solving abilities, determination, and the ability to work with others. A reservoir engineer, for example, may earn nearly $40,000 annually, depending, of course, on education and expertise.

Nonprofessional Jobs

For nonprofessional production jobs, vocational training and/or previous experience are usually recommended. A high school diploma, while not always required, is becoming more important as new technology develops and becomes applicable. Many of these jobs also call for strength, stamina, and the ability to work outdoors in a variety of weather conditions.

Probably the most important worker at a storage site is the pumper. This worker is sometimes called a gauger, although at some companies, the gauger is a person who keeps production records for the well. A pumper operates the equipment used to bring oil and gas to the surface. He or she is ultimately responsible for getting the petroleum from the well to the distribution system or purchaser. This position calls for the strict adherence to standards and regulations set by the company and/or the government.

''The job may not call for rugged physical characteristics, but the pumper or gauger must have a thorough knowledge of the job and equipment, be dependable, and be honest,'' say authors Bill D. Berger and Kenneth E. Anderson in their book *Modern Petroleum: A Basic Primer of the Industry.* A pumper must be able to produce and measure the proper amount of gas and oil from the well and make sure the owner gets proper credit for the amount delivered. A pumper is also responsible for controlling the production from each well. And, in each twenty-four-hour period, the pumper or the gauger must measure the volume of gas, oil, and salt water produced by the well. He or she must also determine the temperature of the oil and measure the tank before and after delivery. Pumpers/gaugers may earn about $20,000 or more annually, depending on training, education, and responsibilities.

The job of a treater is to add chemicals to remove impurities from oil as it is pumped from the well. Switchers are used to regulate the flow of the well when the natural pressure is great enough to force oil from the well without pumping. A lease supervisor is responsible for performing all jobs related to pumping at some wells. When the ownership of the oil is transferred, a pipeline gauger or transport driver makes a final measurement.

Salaries for the above positions are unavailable, but would naturally vary from company to company and, as in other fields, would depend on education and experience.

Tank trucks are used to transport petroleum and finished products such as gasoline, kerosene, and jet fuel. (American Petroleum Institute photo)

CHAPTER 6

TRANSPORTING PETROLEUM

Getting petroleum from the well site to the consumer is a complex process involving many people and several modes of transportation. The basic transportation systems used most by the petroleum industry include pipelines, tankers and barges, highway tank trucks, and railroad tank cars. Each is used for a different purpose and each is designed to move either crude oil or its byproducts in an efficient, economic, and environmentally sound manner.

PIPELINES

Probably the most common—and surely the least visible means of transportation—is the pipeline. Thousands of miles of these lines crisscross the country, delivering millions of barrels of petroleum and billions of cubic feet of natural gas to refineries, processing plants, and consumers. Three kinds of pipelines are used:
- Gathering lines—usually the smallest and used to move oil from producing wells to field storage tanks

- Crude oil trunk lines—used to transport oil from storage tanks to refineries
- Product trunk lines—used to transport refined products to regional distribution centers

One of the most famous pipelines is the 800-mile Trans-Alaska Pipeline System, the largest privately financed construction project in history at a cost of $7.7 billion. A longer, but less costly pipeline, is the 1,200-mile All American Pipeline recently built from California to the Eastern Texas Gulf Coast.

Pipelines under construction on land must follow strict environmental and government regulations. For instance, builders must negotiate with each owner of the land the pipeline will cross. The builder is also responsible for putting together the necessary equipment and crew needed to construct the pipeline. A regular pipeline construction project may require 250 to 300 workers, while a larger pipeline may require 500.

Most pipelines are built underground in a trench dug deep enough to provide an adequate cover. After welders join various lengths of pipe that have been brought to the site, the line is hoisted and laid in the ditch. Dirt is used to cover the line and the area is cleaned up.

To help move oil through pipelines at about three to five miles per hour, pumping stations are established along the route. The distance between each is determined by the terrain, type of oil or product being moved through the line, and the size of the pipe. The flow of the oil is directed by a centrally located operator who controls the pump as well as monitors and controls the flow rate and pumping pressure system. Most systems are now automated and controlled by computer.

Products are moved through pipelines in shipments called batches. To separate batches, an inflatable rubber ball is used. To clean out residue from a previous batch and to keep materials from building up, a scraper called a pig is run through the line. Some pipelines require workers to remove and clean it at each pumping station. It is then replaced and continues through the line.

Other maintenance performed on pipelines include leak detection, which is normally carried out by an aircraft. Regular communications about the pipeline's schedule is also considered maintenance. Companies use private telecommunications systems as well as mobile radios. New technology is allowing companies to integrate computers with microwave systems to improve communications.

Large seagoing vessels used to construct pipelines in offshore waters are called lay barges. One of the largest is called a superbarge, which can house up to 350 workers and store as much as 20,000 tons of pipe.

TANKERS AND BARGES

Using waterways for transporting products is not unique to the petroleum industry. But at no time in the history of the United States were the value of oil tankers realized than during World War II. It is believed that the Allies were victorious because of the tankers' ability to carry large amounts of crude produced in the West to the fighting forces.

Most of the world's petroleum is transported by tankers, which can be as long as 1,300 feet with a carrying capacity of 3.50 million barrels or 147 million gallons. There are over 3,000 vessels in the world's tanker fleet, with average speeds of twelve to fifteen knots. It is those same type of oil tankers flying the flags of their countries that U.S. ships were sent to protect in the Persian Gulf in 1987. Since America imports about forty-six percent of its oil, most of it arrives by tanker.

The United States owns nearly three hundred of its own tankers with a total cargo capacity of more than sixteen million tons. This domestic fleet transports crude between Alaska and refinery centers on the West and Gulf Coasts.

Barges are used extensively on America's 35,000 miles of extensive inland waterway system, including lakes and rivers. This type of vessel is used to transport petroleum products between refining centers and consumers. An average river barge can carry about 11,000 barrels of oil, while the largest ocean-going barge can carry as many as 250,000 barrels.

TANK TRUCKS AND RAIL CARS

Probably the most obvious method of transporting petroleum products is by tank truck, those massive vehicles that can carry loads ranging from 8,000 to 10,000 gallons each. Most modern tank trucks are built of lighter materials, such as aluminum alloys, and run on diesel fuel, which is more economical than gasoline on long hauls. Trucks are also used to transport such finished products as gasoline, kerosene, and jet fuel from the refineries to a customer distribution point.

About 184,000 railroad tank cars are in use today, carrying crude oil, petroleum products, and chemicals. Each is capable of holding between 4,000 and 33,000 gallons. Most railcars are specially designed to handle certain products, whether they need to be kept hot or cold. New materials are also being used to make the rail tank cars more efficient and safer.

EMPLOYMENT OPPORTUNITIES

There are as many opportunities for employment in the transportation segment of the petroleum industry as there are methods of moving the product. Truck drivers, seagoing personnel, railroad workers, and pipeline operators are all needed to get products to their destinations.

Persons interested in a career as a truck driver for petroleum products should have a good driving record. Some states require workers to obtain a commercial motor vehicle operator's license. In general, most truck drivers may have to pass a physical and written exam, as well as a driving test. They should have good hearing, at least 20/40 vision with or without glasses or corrective lenses, be able to lift heavy objects, and be in good health. Gasoline tank truck drivers are responsible for getting to a destination in a timely and safe manner. Once there, the driver attaches the hoses and operates the pumps on their trucks to transfer the gasoline to gas stations' storage tanks.

Trucking companies employed over one-fourth of all truck drivers in 1986. Over one-third of those worked for companies engaged in wholesale and retail, including oil companies.

Other jobs available in the petroleum transportation industry include pipeline construction workers, oil dispatcher, dock supervisor, gaugers, compression station workers, distribution workers, construction and maintenance inspectors, oil pumper, station engineer, barrel filler, gas-transfer operator, and loader. Some of these offer entry-level positions, such as helpers or apprentices.

Salaries for these workers vary. An oil field laborer could earn nearly $14,000 and a pumper/gauger could make as much as $20,000. However, earnings vary from state to state, as well as from company to company. And, as in most positions, the amount of education, training, and experience a person has helps determine that worker's salary.

CHAPTER 7

REFINING AND MARKETING

Before crude oil can be used, it must undergo a process called refining. The plant where the oil is separated into various components and made into the hundreds of products we use each day is called a refinery. Ironically, experiments in refining began even before the petroleum industry was established. Eighteenth-century Europeans attempted to distill coal, shale, and tar from seepages and whale oil in an attempt to develop new products.

Although coal-oil plants were popular in the United States during the mid-1800s, many closed when the petroleum industry mushroomed after the discovery of crude oil in Pennsylvania. The first refinery to open after the Drake discovery was built near Oil Creek in 1860 by William Barnsdall and William H. Abbott. That refinery probably looked a lot different from today's refinery, which appears to be a complex maze of pipes and storage tanks. As modern refining facilities become more and more automated, highly skilled workers are needed to operate them.

There are about two hundred refineries currently operating in the United States. Their refining capacity ranges from the smaller (190 barrels a day) to large-scale plants, which process more than half a million barrels of crude oil daily. Together, these U.S. refineries can process about sixteen million barrels a day. One barrel of oil contains forty-two U.S. gallons and is typically refined into:

- gasoline—eighteen gallons
- kerosene, light fuel oil—ten gallons
- residual fuel oil—five gallons
- jet fuel—three gallons
- lubricating oil, asphalt, wax—two gallons
- chemicals for use in manufacturing (petrochemicals)—two gallons
- other—two gallons

Because there are different types of crude oils, the process of refining is different for each. Crude oils with a paraffin base contain a large amount of paraffin wax, with little or no asphalt. This type of oil yields wax, as well as large amounts of high-grade lubricating oil. As the name implies, asphalt-base crude oil has a large amount of asphaltic materials, while mixed-base crudes have quantities of both paraffin wax and asphalt.

Contrary to popular belief, crude oil is not always a thick, black, gooey substance. Some oils are nearly colorless, while others are amber, green, or brown. Crude oil that has more than one percent sulfur and other mineral impurities are called sour crudes. Those that have less than one percent sulfur are called sweet crudes.

REFINING PROCESSES

All crude oils, no matter what type, must undergo three refining processes: separation, conversion, and treatment.

There are various methods used to separate the various components of the oil. The most common is the distillation process. The crude oil is heated, and as the components vaporize, they are drawn off for further processing. Other ways to separate the oil into parts (also called fractions), include the use of solvents, absorption, and crystallization.

After the crude oil is separated, it must undergo a conversion process. This is where the molecular structure of the separated fractions is changed to produce specific products. This particular process began earlier in this century in response to a growing demand for gasoline. The conversion step lets refiners produce gasoline from groups of hydrocarbons not normally found in the gasoline range.

Processes using heat and pressure and/or chemical catalysts are used to break heavier oils into lighter ones to make such products as gasoline. Similar methods are used to combine several light molecules into a few heavy ones in order to make high-octane fuels.

In the third, and last refining process, oil products are chemically treated to remove impurities and to also improve products.

The natural gas refining business became a thriving industry with the advent of natural gas pipelines. During the conditioning processes, water, impurities, and excess hydrocarbon liquids are removed. Common plant processes include oil absorption, fractionation, dehydration, and cryogenic processing.

MANUFACTURING PETROCHEMICALS

During the refining process, two of the forty-two gallons in each barrel of oil are chemicals used to manufacture other products. These chemicals, known as petrochemicals, are primarily composed of hydrogen, carbon, nitrogen, and sulfur. When the molecules of these ingredients are changed and shifted in various combinations, hundreds of chemical products, called feedstocks, are produced for use in commonly used items.

A petrochemical plant turns petroleum derivatives into feedstocks that will be used to manufacture a wide variety of products used by all Americans on a daily basis. These products include ammonia—as well as polyethylene and polypropylene—which are used to make such items as appliance and automotive parts, luggage, plastic toys, and containers. Other feedstocks are used to make such diverse items as cups and glasses, furniture, fibers to make wearing apparel, surf boards, recording tape, as well as paints.

There are about three hundred petrochemical plants now in operation in the United States. Most are located in what is known as this country's Golden Crescent, a seven-hundred-mile coastal strip between Brownsville, Texas, and New Orleans, Louisiana. This area is noted for its available water transportation, refining complexes, and oil and gas fields.

EMPLOYMENT OPPORTUNITIES IN REFINING

There are a variety of careers available in the refining segment of the petroleum industry. Professional opportunities include those of process and chemical engineers. Process engineers oversee refinery operations, while chemical engineers work in many phases of the production of chemicals and chemical products. Both require college degrees in an appropriate field.

Chemical Engineers

According to the *Occupational Outlook Handbook,* chemical engineers work in many phases of the production of chemicals and chemical products. They may design equipment and plants, determine and test methods of manufacturing the products, and supervise production. Because the duties of chemical engineers cut across many fields, they apply principles of chemistry, physics, mathematics, and mechanical and electrical engineering. Chemical engineers frequently decide to specialize in a particular area, such as oxidation or polymerization. (See chapter 4 for training, advancement, and other information about engineers.)

EMPLOYMENT

The *Handbook* says 52,000 chemical engineers were employed in 1986. Of that number, two-thirds were in manufacturing industries, primarily in the chemical, petroleum refining, and related industries. One-fifth worked for engineering services or consulting firms where they designed chemical plants or did other work on a contract basis. A small number of chemical engineers worked for government agencies or were employed as independent consultants.

JOB OUTLOOK

Employment of chemical engineers is expected to increase fourteen to twenty-four percent through the year 2000, according to the *Occupational Outlook Handbook*. Most openings, however, will result from the need to replace chemical engineers who transfer to other occupations or leave the labor force.

Although output of the chemical industry, where many chemical engineers are employed, is expected to expand, employment of chemical engineers in this industry is not expected to increase. This is due to anticipated productivity improvements, as well as a trend toward contracting out the work. The *Handbook* also notes that the drop in oil prices has reduced opportunities for chemical engineers in petroleum refining and other energy-related industries. Job prospects for chemical engineers working in research on alternative energy sources and energy conservation also hinge on an increase in the price of oil.

Chemical engineers are one of the highest paid in the engineering profession, second only to petroleum engineers. According to the College Placement Council, the average starting salary for a chemical engineer holding a bachelor's degree in 1986 was $29,256.

Other Jobs

The *Occupational Outlook Handbook* reports that in 1986, 31,000 workers were employed in gas and petroleum plant and systems occupations. These refinery positions included:

- Gaugers—gauge and test oil in storage tanks and regulate the flow of oil.
- Petroleum refinery and control panel operators—analyze specifications or follow process schedules to operate and control, using panelboards and continuous petroleum refining and processing units.
- Gas plant operators—distribute or process gas for utility companies and others; distribute the gas for an entire plant or process, often using panelboards, control boards or semiautomatic equipment.
- Petroleum pump systems operators—operate and control manifold and pumping systems to circulate liquids through a petroleum refinery.
- Laboratory analysts—perform extensive tests to maintain quality control on raw materials, process stream, as well as finished products.

A number of nonengineering jobs in the refining industry do not require a college education. Many companies provide classroom and on-the-job training. However, because modern refineries are becoming highly automated, workers should have good mechanical aptitude and knowledge of specialized equipment.

Other jobs associated with the refining field include those in the transportation area, including truck drivers and barge, pipeline, tanker, and railroad workers. Salaries for these workers probably vary from company to company and depend on the amount of experience.

MARKETING

One of the most fascinating aspects of the petroleum industry is marketing, whether trading crude oil on the world market or retailing petroleum products. While the oil market has always had some flux, in recent years it has become more volatile because of the influence of worldwide political events, particularly involving the Middle East.

Planners

Vertically integrated oil companies are large firms involved in all aspects of the oil industry—from exploration to retail sales of its products. Because the company is totally involved in the entire process, marketing plans are integrated within the company's exploration, drilling, production, refining, transportation, and distribution activities.

Oil marketers, whether large or small, must determine the consumer market in order to balance supply and demand. Research must be conducted into present, near future, and distant future demand. Using the material from this research, marketers can determine how well their advertising and public relations programs are working.

A good marketing plan should take into consideration the sources and availability of crude oil, refinery capacity, and ease of shipping and pipelining. Computer programs designed specifically for determining consumer supply and demand can evaluate various options. From this information, a buyer can select the closest possible scenario that would fit a company's marketing plan.

Researchers study government reports, scientific journals, and current news reports to keep abreast of activities in the world oil market that may affect supply and demand. For instance, an explosion at a propylene plant in early 1988 caused a shortage of that petrochemical used to make antifreeze. As a result, the cost of antifreeze rose that winter.

Planners are also responsible for arranging the shipment of products from the refinery to a terminal, either by tanker, pipeline, barge, or railroad. From the terminal, the products are delivered to bulk plants or directly to service stations for use.

Sellers

Independent marketers usually sell a product or products at some point in the marketing chain. These companies purchase name brand refiners' overstock or products from independent refineries. These independent marketers may also be brokers. These are people who buy and sell primarily on paper. Often they purchase products on the spot market; that is, they buy on the spot as the immediate market demands, rather than rely on long-term plans. Independent marketers sometimes have contracts with suppliers, and may also be involved in storage, transportation, and distribution.

Independent middlemen, called jobbers, usually enter into a contract with the owner of the terminal company to sell that product brand wholesale. The jobber could also have his or her own retail store where he or she resells the product from the terminal. Independent service stations may buy their fuels from the company owning the terminal or from jobbers.

Users

Two of the largest users of petroleum products are private homes and commercial buildings. Since crude oil prices rose in the early 1970s, electricity and natural gas have become the top choices for heating and cooling homes and for cooking. Both are also easier to use, since they can be wired or piped to individual facilities. Both natural gas pipelines and electric lines are considered public utilities and therefore subject to government regulations.

The agriculture industry is another large end user of petroleum products. Diesel fuel and liquefied petroleum gas are used for farm equipment, while other petrochemicals are made into fertilizer.

Other types of refined products are used to fuel airplanes, as well as ships.

OPPORTUNITIES IN MARKETING

There are a variety of opportunities available for persons interested in a petroleum marketing career. Administration and management positions are open to persons who have college degrees in related fields, such as business administration or marketing.

In the sales department of a major oil company, sales personnel call on retailers, wholesalers, and commercial, industrial, and agricultural customers to sell the company's products. Persons who exhibit an expertise for directing sales operations do well in managerial sales positions.

In an integrated company, technical support personnel are needed and usually require an engineering degree. Other support staff include clerical workers, customer service representatives, accountants, attorneys, statisticians, data processing specialists, and public relations personnel. Some jobs require advanced training or education, while others offer on-the-job training.

According to the *Occupational Outlook Handbook,* the median annual salary for marketing, advertising, and public relations managers was $35,400 in 1986. For these positions, as well as others in the sales and marketing departments, salary levels vary substantially depending upon the level of managerial responsibility, length of service, and size and location of the firm.

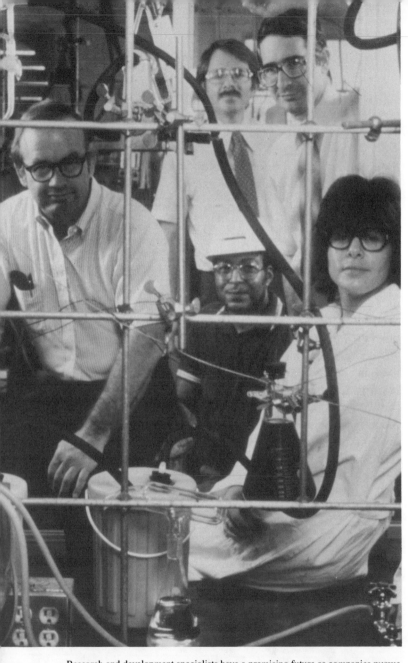

Research and development specialists have a promising future as companies pursue more efficient and economical methods of retrieving petroleum. (Texaco USA photo)

CHAPTER 8

RELATED PETROLEUM OCCUPATIONS

The preceding chapters have covered primary careers in the petroleum industry, as well as touched on some of the most important support service jobs needed to explore, drill, refine, and market refined products. Because the petroleum industry is so complex and covers such a broad area, thousands of workers throughout the world are needed in every aspect of the business. Anyone with marketable skills should be able to find a career in the petroleum industry. Aside from the obvious white-collar, professional jobs available, there is also a need for many types of office workers, as well as blue-collar positions.

RESEARCH JOBS

One area that has somewhat weathered the recent downturn is the research and development field. In fact, many companies have taken advantage of the slowdown to pursue new ideas and methods of retrieving petroleum that are more efficient and economical. During 1988, the term greenhouse effect became a buzzword. But even before this, scientists were seeking ways to protect the environment from oil spills, dust, noise, and refinery emissions. Because the problem became more visible during 1988, environmental scientists will probably intensify their search in an effort to help the petroleum industry become a cleaner business.

The petroleum industry already spends more than $5 billion annually on environmental expenditures, and as new laws and regulations are enacted, the cost is expected to nearly double. Therefore, as the search for new energy reserves and sources continues, the outlook for careers in research and development remains somewhat promising. Some alternative sources currently under study by scientists include the synthetic fuels—also known as synfuels.

Scientists are looking at ways of converting coal into cleaner, more usable petroleum liquids and gases. They are also being challenged into finding a way to economically mine and process oil shale using a minimal amount of water with a minimum amount of environmental damage. While oil can already be extracted from tar sands, scientists continue to search for ways to remove it more economically.

Also high on their list to study are alternative energy sources such as geothermal and solar energy, hydropower, and wind power. With all of these options, it's no wonder that working in research and development would cover a wide variety of fields. However, most workers employed in that particular segment of the industry must have an extensive background in chemistry and other sciences, and usually a Ph.D. in chemical engineering.

Research support staff at a number of major oil companies also includes personnel trained in health and safety engineering, entomology, biochemistry, plant pathology, toxicology, mathematics, and statistics. Another important position is the research scientist, who studies ways of using chemicals to remove oil from coal, shale, and tar sands. They may also study ways to use other energy sources. Research scientists may also be called chemical or technical engineers and may help look for alternate energy technologies, particularly in recovering and refining petroleum products. (See chapter 7 for employment opportunities for chemical engineers.)

CHEMISTS

A career closely related to chemical engineering is that of a chemist. Workers in this field search for and put to practical use new knowledge about chemicals. Throughout the years, they have developed a vast amount of new and improved synthetic fibers, paints, adhesives, drugs, electronic components, lubricants, and other products that use petroleum as their base.

Chemists also develop processes that save energy and reduce pollution, such as improved oil refining methods. A number of chemists work in research and development where a large amount of their time is spent in a laboratory. In basic research, chemists investigate the properties, composition, and structure of matter and the laws that govern the combination of elements and reaction of substances. In applied research and development, they create new products or improve existing ones, often using knowledge gained from their basic research.

Chemists also work in production and inspection of chemical manufacturing plants, where they prepare instructions for workers as to ingredients, mixing times, and temperatures for each stage of the process. They also monitor automated processes to ensure proper product yield, and test samples to ensure they meet industry and government standards. They also record and report on these test results.

Some chemists may work as sales representatives or in the marketing field where they are able to provide technical information on chemical products.

Specialization fields include analytical chemists, who determine the structure, composition, and nature of substances and develop analytical techniques; organic chemists, who study the chemistry of the vast number of carbon compounds and who have developed such products as drugs, plastics, and fertilizers; and, physical chemists, who study the physical characteristics of atoms and molecules and investigate how chemical reactions work. They may also research new and better energy sources.

Working Conditions

Chemists usually work regular hours in offices and laboratories. Some are exposed to health or safety hazards when handling certain chemicals. However, there is little risk if proper procedures are followed.

Employment

Chemists held over 86,000 jobs in 1986, according to the U.S. Department of Labor's *Occupational Outlook Handbook*. Over half of them worked for manufacturing firms, and half of this number worked in the chemical manufacturing industry.

Training, Other Qualifications, and Advancement

A bachelor's degree with a major in chemistry or a related discipline is sufficient for many beginning jobs as a chemist. However, graduate training is required for most research jobs, and most college teaching jobs require a Ph.D.

Beginning chemists with a master's degree can usually teach in a two-year college or go into applied research in government or private industry. A Ph.D. is generally required for basic research, for four-year college faculty positions, or for many administrative positions.

Many colleges and universities offer a bachelor's degree program in chemistry. About 580 are approved by the American Chemical Society. In addition to required courses in analytical, inorganic, organic, and physical chemistry, undergraduates usually study biology, mathematics, physics, and liberal arts.

Students planning careers as chemists should enjoy studying science and mathematics, and should like working with their hands building scientific apparatuses and performing experiments. Perseverance, curiosity, and the ability to concentrate on detail and to work independently are essential.

In government or industry, beginning chemists with a bachelor's degree analyze or test products, work in technical sales or services, or assist senior chemists in research and development laboratories. Employers may have training and orientation programs that provide special knowledge needed for the employer's type of work.

Job Outlook

The *Occupational Outlook Handbook* predicts employment of chemists is expected to increase five to thirteen percent through the year 2000. Like chemical engineers, employment of chemists in the expanding chemical industry is not expected to increase much because of improved productivity methods and contracting. Slow growth also is anticipated in petroleum refining and most other manufacturing industries that employ chemists. However, the openings that will occur will be the result of chemists transferring to other occupations.

Chemistry graduates may become high school teachers, where they would be regarded more as science teachers rather than chemists. Others may qualify as engineers, especially if they have taken engineering courses. Those with a doctorate may become college or university teachers.

Earnings

According to the College Placement Council, chemists with a bachelor's of science degree averaged $23,400 annually in 1986. Those with a master's degree averaged $28,000, while Ph.D.s earned about $36,400. The *Occupational Outlook Handbook* quotes 1986 median salaries from American Chemical Society members as $33,000 for a bachelor's degree; $37,900 for members with a master's degree; and $47,800 for a Ph.D. In a Bureau of Labor Statistics survey, chemists in private industry averaged $22,500 a year in 1986 at the most junior level, and $74,600 at senior supervisory levels. Experienced mid-level chemists with no supervisory responsibilities averaged $41,500.

Depending on a person's college record, the annual starting salary in the federal government in early 1987 for an inexperienced chemist with a bachelor's degree was either $14,822 or $18,358. Those who had two years of graduate study began at $22,458 a year, and with a Ph.D. degree, $27,172 or $32,567. The average salary for all chemists in the federal government in 1986 was $38,600.

Related Occupations

The work of chemical engineers, occupational safety and health workers, agricultural scientists, biological scientists, and chemical technicians is closely related to the work done by chemists. The work of other physical and life science occupations may also be similar to that of chemists.

SCIENCE TECHNICIANS

The *Occupational Outlook Handbook* describes the job of a science technician as one who uses the principles and theories of science and mathematics to solve problems in research and development, production, oil and gas exploration, sales, and customer service. Their jobs are more limited in scope and more practically oriented than those of scientists.

Science technicians who work in research and development construct or maintain experimental equipment, set up and monitor experiments, calculate and record results, and help scientists in other ways. In production, they test products for proper proportions of ingredients or for strength and durability.

Petroleum technicians measure and record physical and geological conditions in oil and gas wells using instruments lowered into wells or by analysis of the mud from wells. In oil and gas exploration, they collect and examine geological data or test geological samples to determine petroleum and mineral content. Some petroleum technicians, called scouts, collect information about oil and gas well-drilling operations, geological and geophysical prospecting, and land or lease contracts.

Working Conditions

Science technicians work under a variety of conditions. Many work indoors, usually in laboratories, and have regular hours. Some occasionally work irregular hours to monitor experiments that can't be completed during regular working hours.

Petroleum technicians perform much of their work outdoors, sometimes in remote locations, and some may be exposed to hazardous conditions. However, there is little risk if proper safety procedures are followed.

Employment

Science technicians held about 227,000 jobs in 1986. About forty percent worked in manufacturing, especially in the chemical, petroleum refining, and food processing industries. Almost forty percent worked in service industries, mainly in colleges and universities, and in independent research and development laboratories. In 1986, the federal government employed about 18,000 science technicians, mostly in the Departments of Defense, Agriculture, Commerce, and Interior.

Training, Other Qualifications, and Advancement

Most employers prefer science technician applicants who have at least two years of specialized training. Many junior and community colleges offer associate degrees in a specific technology or a more general education in science and mathematics.

Technical institutes generally offer technician training, but provide less theory and general education than junior or community colleges. The length of programs at technical institutes varies, although two-year associate degree programs are common.

Many science technicians have a bachelor's degree in science or mathematics, or have had science and math courses in four-year colleges. Some with bachelor's degrees become science technicians because they can't find or don't want a job as a scientist. In some cases, they may be able to move into jobs as scientists, managers, or technical sales workers.

Some companies offer formal or on-the-job training for science technician jobs. Technicians also may qualify for their jobs with some types of armed forces training.

Persons interested in a career as a science technician should take as many high school science and math courses as possible. They should be able to work well with others, since technicians often are part of a team. Technicians usually begin work as trainees in routine positions under the direct supervision of a scientist or experienced technician. As they gain experience, they take on more responsibility and carry out a particular assignment under only general supervision. Some eventually become supervisors.

Job Outlook

The *Occupational Outlook Handbook* says employment of chemical and petroleum technicians is expected to increase five to thirteen percent through the year 2000 due to an expected growth in scientific research and development and production of technical products. Most jobs, however, will be to replace technicians who transfer to other occupations or leave the labor force.

Earnings

Median annual earnings of science technicians in 1986 were about $22,000, according to the *Handbook*. The middle fifty percent earned between $16,200 and $29,400. Ten percent earned less than $12,000 and ten percent earned over $36,000.

In the federal government in 1987, the starting salary for science technicians was $11,802, $13,248, or $14,822, depending on education and experience. The average salary for science technicians employed by the federal government in 1986 was $21,055.

COMPUTER SYSTEMS

Like most industries, the computer age has virtually revolutionized the petroleum business. Most oil companies use computers to keep track of the rapid flow of detailed information. However, major oil firms, as well as drilling companies, are taking advantage of the technical advances being made in computer software to make their operations more efficient.

There are a number of careers open to someone with a degree in computer programming or computer science. Major oil companies usually need professionals in the areas of business; exploration, production, and process; administration; operations; voice and data communications; and technology.

In the business application, a new worker could conceivably begin work as a computer programmer, who is responsible for developing software. Advancement is possible into a coordinating role, working with line management and staff departments to develop and refine data processing applications. The idea here is to develop a comprehensive information management system that keeps track of a company's operations from the wellhead to the refinery to the distributor and to the end user.

Professionals in the computer science field are also needed in most major oil company's exploration and production operations. Nearly all explorationists now use sophisticated computer-generated images to study the earth's formations. Real-time data acquisition, primarily from well logging, is playing an increasing role in the search and production of petroleum. This means additional personnel is needed to develop and maintain software packages to support these necessary functions.

Also opening up are computer careers in the administrative field. Workers in this area are usually responsible for developing, maintaining, and supporting a company's personnel and payroll database. Other functions may include developing a system to forecast, budget, monitor, and report on a wide variety of company activities.

Computer operations specialists must be able to maintain a company's complex data center, as well as update hardware and software as necessary. Electrical engineering, telecommunications, or computer science professionals are needed to plan and install a common system using the latest technology. A computer analyst or systems engineer must also be able to keep up with the latest developments in both hardware and software.

Working Conditions

Computer programmers and computer systems analysts both work in offices in comfortable surroundings. They usually work forty hours a week, as do other professional and office workers. Occasionally, however, evening or weekend work may be necessary to meet deadlines.

Training, Other Qualifications, and Advancement

Persons interested in computer programming may learn the field at public or private vocational schools, community or junior colleges, and universities. Because computers are becoming such an integral part of our everyday life, many high schools are now teaching introductory courses in data processing. Since the field and the number of qualified applicants for new jobs continue to grow, many companies are beginning to require college degrees or higher levels of training.

Employers using computers for scientific or engineering applications prefer college graduates who have degrees in computer or information science, mathematics, engineering, or the physical sciences. Graduate degrees are required for some jobs.

For skilled programmers, the prospects for advancement are good. In large companies, they may be promoted to lead or systems programmers and be given supervisory responsibilities. A programmer may become a systems analyst or be promoted to a managerial position. Because the duties of a systems analyst vary from company to company, there is no universally accepted way of preparing for this career. Most companies, however, prefer college graduates; and for particularly complex tasks, a number of firms seek out workers with graduate degrees.

Employers usually want analysts with a background in accounting or business management for work in a business environment, while a background in the physical sciences, applied mathematics, or engineering is preferred for work in scientifically oriented organizations.

For either the programmer or the analyst, continuous training is essential to keep skills up to date. Certification in either field from the Institute for Certification of Computer Professionals is a sign of experience and professional competence.

Employment and Job Outlook

Systems analysts held about 331,000 jobs in 1986, while programmers held an additional 479,000, according to the *Occupational Outlook Handbook*. As technology becomes increasingly sophisticated and the need for experienced computer professionals grows, the outlook is for a thirty-five percent growth in the field through the year 2000. Those workers with the best education, training, and up-to-date knowledge, should be top candidates for jobs in the computer field as it relates to the petroleum industry.

Earnings

The *Handbook* notes that 1986 median earnings of full-time programmers were about $27,000 a year. For that same year, median salaries for systems analysts were about $32,800. The *Handbook* also notes that for both programmers and analysts, salaries are highest in the fields of mining and public utilities.

BUSINESS

As in any other major industry, the petroleum field requires a vast number of business professionals to help keep the company running. Applicants with a background in finance, preferably those with a master's of business administration, may work with the company's comptroller. Their duties may include working with cash, asset and liability management, as well as financial planning, forecasting, and economic analysis. They may also be involved in structuring the financing of large capital projects, a common occurrence in the petroleum business. As part of the finance team at a large petroleum company, a worker will probably have to deal with bank representatives, as well as other professionals from companies that may be participating in a joint venture.

Accounting professionals in a petroleum company are responsible for preparing finance statements and reports, conducting audits, and maintaining internal financial controls. Other duties may include monitoring and assessing the ongoing performance, as well as working on financial and economic planning and analysis. Most workers in this department become specialized in any of a number of areas, resulting in the promotion to another facility or location for general managerial responsibilities.

Another career available to business or industrial and labor relations graduates with advanced degrees is human resources. This field covers a wide range of duties, including employment, compensation and benefits, labor relations, career development, and human resource planning.

A promotion to human resource representative would entail working with employees in their day-to-day relations with the company or becoming involved in labor relations. A human resource representative may also be responsible for implementing programs designed to promote career and management development.

Public relations is another career option for persons who write well and enjoy working with a wide variety of people. The director of public relations at a petroleum company is responsible for letting the residents of a community know what the company is doing. Press releases, which contain information about the company's projects, are written by personnel in the public relations department and are sent to newspapers, television, and radio stations, as well as magazines. Press releases may cover a wide variety of topics, including personnel changes in the company, a new discovery, or the company's financial status. Public relations department personnel are also responsible for gathering information from employees to include in a company newsletter.

In most companies, workers in this department must have at least an undergraduate degree in English or journalism. Applicants should also have some hands-on experience working in the media and have a working knowledge of how each is operated.

GENERAL CAREERS

One unique career opportunity in the petroleum industry is the food preparation business. Drilling companies are usually responsible for hiring a catering company that specializes in feeding rig workers, particularly offshore where menus must be planned in advance and supplies brought in by boat or helicopter. Working for a catering company is ideal for someone who likes to cook, but would like a career out of the traditional restaurant or industrial setting.

Petroleum-related companies also employ secretaries, clerks, warehouse personnel, truck drivers, and laborers, as well as welders and mechanics. However, because of the general nature of these jobs, there is probably less opportunity in these types of careers than in a specialized area of the industry. In these type of general career opportunities, salaries and training will vary from company to company.

For general information on these types of jobs, check the *Occupational Outlook Handbook* and the *Dictionary of Occupational Titles*. Both are published by the U.S. Department of Labor and may be found at a city or county library, or at a state Department of Labor's Job Service Office.

Employees in the petroleum industry enjoy excellent salaries, benefits, and a work schedule that allows them to continue their education. (*top:* photo by Dennis Sullivan, Lafayette, Louisiana; *bottom:* American Petroleum Institute photo)

CHAPTER 9

EMPLOYMENT OUTLOOK FOR THE FUTURE

No other industry in the world has been under such close scrutiny and affected by world politics like the oil industry has been for the past fifteen years. Prior to 1973, when the Arabs cut off the supply of petroleum coming into the United States, no one was aware of just how strong a hold these Middle East countries had on the economic survival of the United States. But as Americans lined up to buy fuel for their cars, and other petroleum products became in short supply, they quickly found out exactly what dependence on foreign sources meant. It was a lesson that was repeated on a smaller scale in the late 1970s and one that was soon forgotten.

In January of 1989, the United States was importing forty-six percent of all the oil it uses. Only three years ago, imports stood at thirty-one percent. The highest level of imports ever was forty-eight percent in 1979, the year of the second oil crisis.

In other words, almost half of all the crude oil used by Americans today is not produced in the United States but comes from countries throughout the world.

There has been much debate in the past few years over the United States's growing dependence on imports, in particular from unstable Middle East countries belonging to the Organization of Petroleum Exporting Countries (OPEC). These countries, in recent years, are said to have cheated on allotted production quotas, causing an over-supply on the world market and, as a result, dramatically lower oil prices. In 1986, the price of oil collapsed to the $10 per barrel range. Because of this, U.S. companies have scaled back exploration and production efforts because they are economically unfeasible. They also find it hard to compete with foreign oil companies, which are owned and backed by their respective governments.

From 1986 to 1987, crude oil production dropped from 3.17 billion barrels to 3.05 billion barrels. During the same time, however, energy consumption rose from 74.2 to 76.3 quadrillion BTUs (British thermal units).

OIL AND GOVERNMENT

The oil and gas industry plays a large role in funding governments at the federal, state, and local levels. From 1980 until 1988, when a windfall profits tax on the industry was repealed, energy companies paid twice the amount of federal taxes than any other industry. In 1987, the petroleum industry paid over $3.65 billion to state and local governments in the form of severance and production taxes.

The oil industry is also playing a vital role in filling the U.S. government's Strategic Petroleum Reserve. The goal of the SPR, authorized by Congress in December 1975, is to stockpile one billion barrels of petroleum products in case supplies are disrupted, as they were during the first oil crisis. As of January 1989, about 550 million barrels of oil had been stored in six underground storage facilities in salt domes along the Texas and Louisiana Gulf of Mexico coasts.

But when it comes to drilling new wells, companies say $15 per barrel of oil is too low to make it worth their while. Many of them in the past few years have found it is more economical to buy their reserves through mergers and acquisitions, rather than drilling and exploration. As a result, companies have had to institute a number of cost-cutting measures, including massive staff reductions through either layoffs or early retirements. Most companies, however, say they would drill new wells if oil prices reach—and, more importantly, stabilize—in the $18 to $20 per barrel range.

It's estimated that over 334,000 petroleum industry workers lost their jobs between 1981, the peak of the boom, through 1987. At the end of 1987, about 1.5 million workers were employed in all phases of the petroleum industry. If history continues to repeat itself, the petroleum industry's bust cycle should be broken by the time 1990 college freshmen complete their studies.

As of this writing, all signals point to higher oil prices during the 1990s. And until alternate fuels are developed, oil and natural gas will remain the number one energy source well into the next century.

At the end of 1987, it was estimated that world crude oil reserves stood at 887.3 billion barrels. Of that amount, 699 billion barrels, or 70.5 percent, are located in six OPEC countries. In Saudi Arabia alone, reserves are estimated at 255 billion, or one-fourth, of the world's known reserves. By comparison, the United States has estimated oil reserves of twenty-seven billion barrels and the Soviet Union, the world's largest oil producer, has about 63 billion barrels.

Natural gas, in oversupply for the past several years, is also making a comeback, in part because it burns cleaner and because of the abundant supplies now believed to exist. Companies, as well as governments, also like natural gas because its price is not influenced by OPEC. The industry, however, is still heavily regulated by laws administered by the Federal Energy Regulatory Commission.

The 1987 price for natural gas at the wellhead was $1.71 per thousand cubic feet (mcf). Although that's still lower than the $1.94 mcf companies realized in 1986, prices, while still low, were approaching the $2 per mcf range at the beginning of 1988.

IS A CAREER IN PETROLEUM FOR YOU?

The petroleum industry can provide a rewarding career, as well as the chance for a more than comfortable lifestyle. Salaries are competitive, if not higher, than those in other

U.S. industries. According to the Bureau of Labor Statistics (BLS), workers in the oil and gas industry earned an average of $14.02 an hour in 1987. And even in a depressed industry, it was up from the $13.75 recorded in 1986 and was the highest in the industry's history. The BLS reports that workers in all other manufacturing industries earn an average of $9.91 an hour.

Most companies offer liberal benefit packages, including paid vacations and holidays. Most provide at least part of an insurance program, including all or some life, health, disability, travel, and accident. Other substantial benefits include pension plans, as well as stock bonus plans, which let employees share in the company's success.

Working conditions vary from job to job, as well as location. Personnel in the exploration, drilling, and production divisions perform physically demanding work in a variety of settings and climates. They may also work irregular shifts and travel frequently to perform duties associated with their jobs.

Some petroleum companies, or support service firms, offer on-the-job training, while others encourage additional education courses. Advancement is offered to those who are motivated and willing to learn.

If you plan on making the petroleum industry a lifelong career, keep in mind the cyclical nature of the business. A personality able to cope with the industry's ups and downs is crucial to survival, as is being financially prepared to weather the boom-to-bust cycle.

PREPARING FOR A FUTURE IN THE
PETROLEUM INDUSTRY

If you're still in high school, a good place to begin is with your guidance counselor. Your skills and aptitudes should have already been analyzed through a variety of standard examinations. These should give you a clear idea of the area in which you already excel. For any career route you take, don't forget that the basics of reading, writing, and arithmetic are important. Most companies today require that employees have a high school diploma or GED equivalent.

If you're leaning towards a professional career in the petroleum industry that requires a college degree, you should study courses in high school that will prepare you for a tougher college curriculum. For instance, future engineers should take as many science and math courses as possible. Any type of course that would help familiarize you with computers is also highly recommended.

You should also begin to inquire about entrance requirements to colleges and universities that specialize in petroleum-related fields. Be sure it is in an area of the country that you will be happy living in and one that you can afford.

If money is a problem, start inquiring about scholarships, grants, and other sources of funding for your education. Many local chapters of national and international petroleum organizations and associations award scholarships to high school graduates, as well as college freshmen in a particular petroleum discipline. For a list of cities with local chapters, write to those organizations listed in appendix C.

If you're interested in learning a specific trade, there are a number of both private and public vocational-technical schools that offer a variety of courses that would be applicable to a career in the petroleum industry. Whether you enter college or a trade school, you should make an effort to find out as much as you can about the school. Try to talk to a recent graduate and learn about the school's reputation in that particular field.

If you're lucky, you can sometimes get a job in the petroleum industry that provides on-the-job training. This is the way a number of companies hire various types of personnel, including laborers, clerks, and secretaries, for example. Many of them also provide in-house training for a particular job, and opportunities for advancement.

Adults who want to change careers but cannot afford to leave their present job may want to consider taking evening courses through an adult education program. Many cities offer some type of weekend or night program for adults. Others living in a college town may want to consider obtaining another degree or completing a degree.

In some oil towns, colleges and trade schools design curriculum around an oil field worker's seven and seven schedule. You may want to consider taking a job that would cater to this schedule. This would enable you to continue working while furthering your education at the same time.

Contact others in your field of interest and ask them to recommend training programs. Other sources of information include trade journals, the commercial listings in your telephone book, and local and state boards of education.

LOOKING FOR A JOB

One of the most crucial tools needed in your job search is a résumé. It is a must for anyone who is seeking employment, whether after college, after training, or right out of high school.

There is much debate about how long a résumé should be, whether it should include personal information, or if references should be included. The general consensus is that a résumé should be no longer than one page, should include a work history that lists your most current position first, and should highlight your accomplishments without going overboard. References should be listed on a separate sheet and submitted only when requested. Equal opportunity laws make it illegal for a prospective employer to ask a person their age, sex, race, religion, or marital status; therefore, personal information is optional. But be sure to include your name, address, and telephone number.

Don't print your résumé on colored paper in order to be noticed. You may give the wrong kind of impression. Use only white, off-white, or buff-colored paper for a more professional, polished look.

If you're about to complete your college studies, one of the best places to start is the placement office at your university. Because oil companies use a variety of personnel in their daily operations, many like to recruit recent college graduates. Oil companies contact placement offices and make a date for the company recruiter to visit the campus. So, if you're still in school, check periodically with the placement office. Some offices post the notices of the company recruiter's visit on highly visible bulletin boards near the placement office. Others are posted in heavily trafficked areas throughout campus, such as student unions or buildings where most courses associated with a curriculum are taught. You probably will need to schedule an appointment with the recruiter prior to his or her visit.

Even if you're out of school, you may be able to register with the placement office. A number of schools let their graduates complete an information card, which is then matched to companies seeking employees. Also, keep in touch with your curriculum advisors. They may come across opportunities not initially available to you.

Another good job-hunting source is local chapters of trade organizations or professional societies. Members are in touch with others in the industry and may hear of openings that would fit your education and training. Some associations have student chapters on campus.

The classified section of the daily newspaper is another option for job seekers. Many companies advertise in the state's largest newspaper, or in the publication closest to the area's oil patch. Be prepared to answer blind ads. These are advertisements that describe a position, but give a box number as an address. Many companies use this method in order to screen applicants, as well as avoid a deluge of phone calls and résumés from those who may not qualify for the position.

Employment agencies sometimes offer jobs in the petroleum field. However, they charge finder's fees that are sometimes paid by the employer; otherwise, the fee must be paid by the worker once he or she is hired. Be wary of agencies that request an up-front fee.

A head-hunting agency is usually hired by an employer seeking a special type of person in a particular field. This worker is usually already employed and the agency attempts to woo him or her away. Fees to head-hunting agencies are almost always paid by the employer.

The U.S. Department of Labor and state employment services have offices in major cities. This may be another option when job hunting. Check on other city or county services that may be offered in your particular area.

Other ways of getting information on jobs include talking to a recognized expert in the field or a management consultant, the local Chamber of Commerce, and others who have recently graduated in your curriculum and have found jobs. Also, attend petroleum exhibits and conferences that attract top executives from the industry. One of the largest shows is the Offshore Technology Conference in Houston, Texas, which attracts hundreds of oil company exhibitors from all over the world. Thousands of visitors attend the conference, which is usually held each year in early May. In Lafayette, Louisiana, the Louisiana Gulf Coast Oil Exposition is held in odd-numbered years and features new technology. A similar show is held in Odessa, Texas, in even-numbered years. Other shows are held periodically throughout the United States and the world.

Use the telephone. While it is a good idea to contact company personnel departments, don't forget to call the department heads of companies you are interested in working for and ask for an interview. These people usually know before the personnel department whether an opening is anticipated. Be brief, but to the point. Remember, these people are busy. Do your homework and have a short statement prepared on how you can help the company. If talking to the personnel department, ask about sending your résumé to keep on file. Also, ask them about a summer job or internship program.

Mailing a résumé to the personnel department and department head is a good idea, although not usually as effective as the personal phone call. It's usually more effective when preceded by a personal phone call. In any case, whether responding to an ad for a specific job, or taking a shot in the dark with a department head, a good cover letter is a must. It serves as your personal introduction to a prospective employer. Make it interesting, but to the point. Do not write a generic letter for use in any situation. Write a separate letter specially tailored for a specific job. In closing, be sure to ask for an interview at the employer's convenience.

INTERVIEWING

One of the most difficult parts of the job-hunting process is the interview. This is where you can sell yourself to a prospective employer, so it is essential to be prepared.

First, decide in advance what you are going to wear to the interview. Make sure you are dressed properly, your clothes are neat and clean, and your shoes are polished. Get a good night's sleep and leave early enough to allow for the unexpected, such as car trouble or excessive traffic. Be sure you have the exact name and title of the person you are going to interview with. And take note that it's important to make a good impression with the receptionist, as well as any other employee you might meet during your visit to the company.

Investigate the company before the interview. Learn the company's products and services, the competition, the problems. The city or county library is a good place to start. If possible, call the daily newspaper's business editor or someone in that department to find out more information. Also, be sure you understand the job for which you are applying. If necessary, call in advance for more details if you are unsure about any aspect of the position.

Most interviewers usually begin an interview by asking applicants to give a brief description of themselves. Keep the personal aspect brief. Instead, use this opportunity to highlight your professional achievements, particularly as they relate to the position for which you're applying. Be prepared to tell the interviewer about personal accomplishments as well as explain gaps in your work history and short employment periods. In advance of the interview, work on answers that might turn any negatives into positives.

One of the trickiest parts of an interview is the discussion of pay. It is important to know the salary range in advance and be prepared to give an answer you can both live with.

An applicant should also have a list of questions to ask the interviewer. This is a good indication that the applicant has done his or her homework and is truly interested in the position. Some questions could include asking about the person who previously held the job, daily duties, working conditions, and chances for advancement. The most important question to ask, however, is for the job. If you feel the interview has gone well and you're truly interested, make it clear to the interviewer that you would like the position.

Before you leave, be sure to thank the interviewer for his or her time, while giving a strong handshake.

A handwritten note, either mailed the next day or hand delivered, is also appropriate. It should thank the interviewer for his or her time, briefly recap your conversation, and stress your desire and ability to fit the position.

For more details on résumé-writing and interviewing techniques, check your city or county library for any of the many books written on the art of job hunting.

CONCLUSION

While low-priced energy may bode well for some parts of the country, it has created economic chaos in the country's Oil Belt and, some experts warn, put the United States at risk in case of a national emergency.

In mid-January of 1989, the American Petroleum Institute, one of the industry's main trade organizations, issued a report that warned of slowing U.S. crude oil production. API calculates that U.S. production in 1988 was 8.1 million barrels a day—the lowest in twelve years and a three percent decline from the previous year. In 1985, production stood at 9.1 million barrels a day. The number of rigs exploring for oil also dropped under the 1,000 mark, while the amount of federal lands under lease has fallen from more than 140 million acres in 1983 to less than half that today.

The U.S. energy situation became an issue in the 1988 presidential race between then Vice President George Bush and his Democratic opponent, Massachusetts Governor

Michael Dukakis. Bush, a Texan and former oil man, is said to be a friend of the petroleum business, while Dukakis heads a state that uses large amounts of energy. The polarity between the two, if nothing else, gave oil people a forum through which the nation's lack of an energy policy could be discussed.

How much of the warning will be heeded is anyone's guess. But one thing is for sure: Since Drake discovered oil in Pennsylvania 130 years ago, the petroleum industry has continuously provided a challenge to those who seek and use it. And, in today's economic climate, there's no other business quite like it.

GLOSSARY OF PETROLEUM TERMINOLOGY

Workers in the petroleum industry speak their own language. While many more words and phrases are used in the everyday business of searching for petroleum than are listed here, this oil field dictionary is designed to give you a technical description of the industry. The listings are from a booklet entitled *The Language of Energy: A Glossary of Words and Phrases Used in the Energy Industry.* It is published by the American Petroleum Institute in Washington, D.C.

Not all of the words listed here are included in the preceding chapters. For more details on any of the subjects, check your local library or contact the appropriate organization listed in appendix C.

Acid rain. Acidity in rain or snow produced when carbon, nitrogen, and sulfur compounds oxidize in the atmosphere.

Active solar. A system in which mechanical or electrical devices are used to transform solar energy into heat for space heating or other useful energy products. Contrasts with passive solar.

Allocation controls. Government policy which specifies quantities of goods or services that potential customers may purchase. May also designate specific suppliers for such purchases. Crude oil and petroleum products were subject to federal allocation controls through most of the 1970s.

Anthracite coal. A variety of coal with a high heat content (22–28 million Btu per ton). U.S. production is primarily in northeastern Pennsylvania. It was once prized for home heating due to its low content of ash and other impurities. Also called hard coal.

Aquifer. Underground water reservoir contained between layers of rock, sand, or gravel.

Arab Oil Embargo of 1973–74. During the Arab–Israeli War in October 1973, Arab oil-producing nations agreed to cut off oil shipments to the United States and the Netherlands because they supported Israel. Arab producers simultaneously reduced output. In practice, the shortfall was spread among all oil-importing nations. World prices moved sharply higher. Price and allocation controls suppressed some of the decrease in the United States, but nevertheless led to gasoline lines.

Areas of Critical Environmental Concern. Areas designated by the U.S. Department of the Interior to have historic, cultural, scenic, or natural value that requires protection.

Backfitting (or retrofitting). Modifying equipment to make changes or add features that have been included in later models.

Baghouse. An air filtering device that removes particulates—such as carbon or catalyst fines—from furnaces and process unit exhausts.

Balance of payments. A tabulation of a nation's transactions with the rest of the world that shows the extent to which domestic goods, services, and assets have been transferred to foreign countries and vice versa.

Balance of trade. The difference between receipts from foreigners for a nation's goods and services and payments to foreigners for imported goods and services.

Barrel. The standard measurement in the oil industry. A barrel of oil equals forty-two U.S. gallons. The measurement originates from the wooden barrels used to transport oil in the early days of oil production.

Basin. A depression in the earth in which sedimentary materials have accumulated over a long period of time. A basin may contain many oil or gas fields.

Bcf. Billion cubic feet. The cubic foot is a standard unit of measure for quantities of gas at atmospheric pressure.

Benchmark price. A benchmark is a standard by which things may be measured. A frequently used benchmark price of oil is the price set by OPEC for Arabian Light crude oil, a 34-degree gravity oil produced in Saudi Arabia. The other OPEC nations set their official prices in accordance with agreed differentials from the benchmark price. In the United States, the most commonly used is West Texas Intermediate.

Best Available Control Technology (BACT). The maximum degree of emissions control that a permitting authority determines to be achievable, considering energy, environmental, and economic impact as well as other costs.

Biomass. Any kind of organic substance that can be turned into fuel, such as wood, dry plants, and organic wastes.

Bituminous coal. A variety of coal with high heat content (19–30 million Btu per ton) soft enough to be easily ground for combustion. The most widely mined and consumed coal. Also called soft coal.

Bonus payment. Cash paid to a landowner or other holder of mineral rights by the successful bidder on a mineral lease in addition to any rental or royalty obligations specified in the lease. Most sales of oil and gas leases on federal lands (on or offshore) involve bonus payments.

Btu. British thermal unit. A standard measure of heat content in a substance that can be burned to provide energy, such as oil, gas, or coal. One Btu equals the amount of energy required to raise the temperature of one pound of water one degree Fahrenheit at or near 39.2 degrees Fahrenheit.

Carbon monoxide emissions. Colorless, odorless toxic gas sent into the air when carbon molecules are burned incompletely. Nearly all forms of fossil fuel combustion emit carbon oxides.

Cartel. Originally a combination of commercial enterprises that formally agree to limit competition by setting prices (often accompanied by output quotas). The term now refers to any group of enterprises that are suspected or accused of anticompetitive behavior. The Organization of Petroleum Exporting Countries (OPEC) has attempted to act as a cartel in the original sense.

Coal gasification. The chemical conversion of coal to synthetic gaseous fuels.

Coal liquefaction. The chemical conversion of coal to synthetic liquid fuels. (See also direct and indirect coal liquefaction.)

Coal slurry pipeline. A pipeline used to transport coal long distances by mixing crushed coal with water and sending the liquid mixture (or slurry) through pipelines.

Coal washing. Cleaning coal with water and certain additives before burning to remove some of the impurities. Most impurities tend to be heavier than coal and sink to the bottom of the water mixture so the clean coal can be floated off.

Cogeneration. The combined production of electrical or mechanical energy and usable heat energy.

Completed well. A well made ready to produce oil or natural gas. Completion involves cleaning out the well, running steel casing and tubing in the hole, adding permanent surface control equipment, and perforating the casing so oil or gas can flow into the well and be brought to the surface.

Conventional sources of energy. Usually refers to oil, gas, and coal (and sometimes nuclear power) as contrasted with alternative energy sources such as solar, hydro, and geothermal power, synfuels, and various forms of biomass energy.

Core. Samples of subsurface rocks taken as a well is being drilled.

Crude oil equivalent. A measure of energy content that converts units of different kinds of energy into the energy equivalent of barrels of oil.

Cushion. In discussions of natural gas, refers to the effects of a supply of low-cost, price-controlled gas. The

Natural Gas Policy Act of 1978 set up an array of categories of natural gas, some of which are controlled at very low prices. A pipeline that has access to large amounts of this low-cost gas is said to have a deep cushion. This cushion could be used to pay above-market prices for other categories of gas and still enable the pipeline to sell gas at prevailing market prices.

Decontrol. Removing government controls on prices and other factors of market activity. Price controls on crude oil began in 1971 and were phased out beginning in 1979. Controls on crude oil ended on January 28, 1981. Price controls on natural gas remain in effect.

Deepwater port. A marine terminal constructed off-shore to accommodate large vessels, in particular large tankers. The terminal is connected to the shore by sub-merged pipelines.

Demand. The quantity of goods or services that an individual or group wants to buy at a given price.

Development well. A well drilled in an already dis-covered oil or gas field.

Direct coal liquefaction. A process by which liquid fuels are produced from the interaction of coal and hydrogen at high temperature and pressure.

Distillate. A generic term for several petroleum fuels that are heavier than gasoline and lighter than residual fuels. Home heating oil, diesel, and jet fuels are the most common types of distillate fuels.

Distributor. A wholesaler of gasoline and other petroleum products. Also known as a jobber. For natural gas, the distributor is almost always a regulated utility company.

Domestic production. Oil and gas produced in the United States in contrast to imported supplies.

Downstream. All operations that take place after crude oil is produced, including transportation, refining, and marketing.

Drill bit. The part of the drilling tool that cuts through rock strata.

Drill string. Lengths of steel tubing screwed together to form a pipe connecting the drill bit to the drilling rig. The string is rotated to drill the hole and also serves as a conduit for drilling mud. Also called the drill pipe or drill stem.

Drilling mud. An emulsion of water, clays, chemical additives, and weighting materials that flushes rock cuttings from a well, lubricates, and cools the drill bit, and maintains the required pressure at the bottom of the well. Examination of cuttings returned to the surface helps geologists evaluate underground rock formations.

Drilling rig. The surface equipment used to drill for oil or gas consisting chiefly of a derrick, a winch for lifting and lowering drill pipe, a rotary table to turn the drill pipe, and engines to drive the winch and rotary table.

Dry hole. A well that either finds no oil or gas, or finds too little to make it financially worthwhile to produce.

Dry natural gas. Natural gas containing few or no natural gas liquids (liquid petroleum mixed with gas).

Economic efficiency. The absence of waste. If a given commodity can be produced by two techniques, the technique that minimizes cost (an aggregate measure of resources used) per unit of output is more efficient.

Effluent. Any waste liquid, gas, or vapor that results from industrial processes. May be harmful to the environment unless properly treated.

Electrostatic precipitator. A pollution control device that removes particulates from industrial stack emissions, thereby preventing or greatly reducing their discharge into the atmosphere.

Embargo. A government order prohibiting commerce. Most frequently a prohibition on exports or imports of a given commodity to or from a nation or nations. (See Arab oil embargo.)

Emissions. Gases and particulates discharged into the environment, usually the atmosphere.

End-use restrictions. A legal prohibition on the use of a commodity. The Powerplant and Industrial Fuel Use Act of 1978 prohibited utilities and large industrial consumers from using natural gas as a boiler fuel in new installations.

Energy-GNP ratio. The amount of energy used to produce a dollar's worth of output, as measured by the gross national product or GNP.

Enhanced oil recovery. Injection of water, steam, gases, or chemicals into underground reservoirs to cause oil to flow toward producing wells, thus permitting more recovery than would have been possible from natural pressure or pumping alone.

Environmental Impact Statement (EIS). A statement of the anticipated effect of a particular action on the environment. An EIS is required of federal agencies by the National Environmental Policy Act of 1969 for any significant action (including granting certain oil and gas permits).

Equilibrium. A market situation in which the quantities supplied by sellers match the quantity demanded by buyers at the current price.

ESECA. Energy Supply and Environmental Coordination Act, passed in 1974, which prohibited certain power plants with coal-burning capabilities from burning petroleum and required newly constructed fossil fuel boilers to be designed to burn coal.

Ethanol. The two-carbon atom alcohol present in the greatest proportion upon the fermentation of grain and other renewable resources such as potatoes, sugar, or timber. Also called grain alcohol.

Excise tax. A tax levied on the production, sale, or consumption of a commodity.

Exploratory well. A well drilled to an unexplored depth or in unproven territory, either in search of a new reservoir or to extend the known limits of a field that is already partly developed.

Exxon Donor Solvent (EDS) Process. A direct coal liquefaction process that dissolves coal in a solvent while adding hydrogen and subjecting it to heat and pressure.

Fault. A displacement of subsurface layers of earth or rock that sometimes seals an oil-bearing formation by placing it next to a nonporous formation.

Federal lands. See government or public lands.

Field. A geographical area under which one or more oil or gas reservoirs lie, all of them related to the same geological structure.

Five–year offshore leasing program. The first step in the process of leasing offshore lands for oil and gas exploration. The Department of the Interior publishes and annually updates a five-year plan of timetables and areas that will be offered for lease.

Fixed bed gasification. A gasification process in which the raw material is fed as uniform-sized lumps and in which the gas moves through a nearly stationary bed of reacting fuel.

Fluidized bed. A bed of fine particles through which a fluid is passed with a velocity high enough for the solid particles to separate and become freely supported in the fluid.

Fly up. The possibility of a sudden, sharp increase in prices of natural gas decontrolled in 1985 under the Natural Gas Policy Act of 1978.

Fossil fuels. Fuels that originated from the remains of plant, animal, and sea life of previous geological areas. Crude oil, natural gas, coal, shale oil, tar sands, lignite, and peat are fossil fuels.

FUA. Powerplant and Industrial Fuel Use Act, passed in 1978, which extended the provisions of ESECA and set specific goals for replacing oil- and gas-burning boilers with coal-fired power.

Geopressured brine. Salt water, unusually hot and in some instances saturated with methane, contained under abnormally high pressure in some sedimentary rocks.

Geothermal energy. Energy produced from heat deep in the earth, usually caused by underground water being heated as it flows through hot underground rocks.

Glut. An excess of supply over demand. In a market system, a glut of a product would cause producers' inventories to increase. The producers would lower the price to bring down their inventories until supply and demand were equal.

GNP. Gross National Product. Total value at market prices of all goods and services produced by the nation's economy. Used as a measure of economic activity.

Government lands. Lands owned by the federal government. (See public lands.)

Greenhouse effect. The theory that increasing concentrations of carbon dioxide in the atmosphere trap additional heat and moisture and can, in time, create a hothouse effect, raising the temperature of the earth.

Groundwater. Water in underground rock strata that supplies wells and springs.

H-Coal Process. A direct coal liquefaction process that adds hydrogen and a catalyst to a coal slurry in a liquefaction vessel.

Heavy oil. Crude petroleum characterized by high viscosity and a high carbon-to-hydrogen ratio. It is often difficult to produce heavy oil by conventional techniques, so more costly methods must be used.

Heavy oil sands. Rocks containing viscous hydrocarbons (other than coal, oil shale, or tar) that barely flow at reservoir conditions. Hydrocarbons in heavy oil sands flow more readily than those in tar sands but much less readily than those in lighter oil sands.

Hopper car. A freight car designed to carry bulk materials such as coal or grain. Its floor slopes to one or more hinged doors for rapid discharge.

Hydrocarbons. Any of a large class of organic compounds containing only carbon and hydrogen. The molecular structure of hydrocarbon compounds varies from the simplest, methane (CH_4), to heavier and more complex molecules, such as octane (C_8H_{18}), a constituent of crude oil. Crude oil and natural gas are often referred to as hydrocarbons or hydrocarbon fuels.

Hydroelectric power. Electric energy produced by harnessing falling water. In most cases, river water flowing through a dam turns a turbine that generates electricity.

Incremental pricing. A provision of the Natural Gas Policy Act of 1978 that requires natural gas price increases to be charged to large industrial users instead of residential users, until the price of gas reaches the price of an alternative fuel.

Indirect coal liquefaction. A method of making synfuels in which coal is first converted to synthetic gas, then catalyzed to produce hydrocarbons or methanol. Additional processing can convert methanol to gasoline.

Interstate pipeline. A pipeline carrying oil or natural gas across state lines. Interstate pipelines are regulated by the federal government.

Intrastate natural gas market. The market for natural gas produced and sold for delivery within a state, as opposed to sales across state lines. Intrastate markets had particular significance prior to the passage of the Natural Gas Policy Act of 1978. Until that time, they were not regulated by the federal government, so prices reflected supply and demand rather than federal regulators' decisions.

In situ. In its original place. Refers to methods of producing synfuels underground (underground gasification of a coal seam or heating oil shale underground to release the oil).

Iranian cutoff. Sharp reductions of Iranian oil supplies in world markets due to the events surrounding the 1979 Iranian revolution.

Kerogen. The hydrocarbon in oil shales.

Lease offering. Also called lease sale. An area of land offered for lease usually by the U.S. Department of the Interior for the exploration for and production of specific natural resources, such as oil and gas. An oil or gas lease conveys no title or occupancy rights, apart from the right to look for and produce petroleum subject to the conditions stated in the lease.

Lignite (or lignite coal). A solid fuel of a higher grade than peat but a lower grade than bituminous coal. Lignite has a high content of moisture and volatile gases. Thus it is soft and has a relatively low heat content, at most 8,300 Btu per pound.

Liquefied natural gas (LNG). Natural gas that has been converted to a liquid by reducing its temperature to minus 260 degrees Fahrenheit at atmospheric pressure. Gas shipped by seagoing tankers is liquefied before being pumped into the ships and is regasified for pipeline transportation upon reaching its destination.

Liquefied petroleum gases (LPG). Hydrocarbon fractions lighter than gasoline, such as ethane, propane, and butane. They are kept in a liquid state through compression and/or refrigeration and are marketed for various industrial and domestic gas uses. Commonly referred to as bottled gas.

Market. A context in which goods are bought and sold, not necessarily confined to a particular geographic location.

Market forces. Pressures produced by the free play of supply and demand in a competitive market that induce adjustments in prices and quantities sold.

Market system. An economy that relies on market forces to allocate scarce resources, determine production techniques, and price and distribute goods and services. Also referred to as a price system.

Mbde. Million barrels a day of oil equivalent.

Metallurgical coal. A quality of coal used for making coke and steel. Low in ash and sulfur and strong enough to withstand handling. Usually composed of a blend of two grades of bituminous coal. Also called coking coal.

Methanation. The final step in high-Btu gas production in which hydrogen-rich gas reacts with carbon monoxide in the presence of a catalyst to form methane.

Methane. A light, odorless, flammable gas that is the chief constituent of natural gas.

Methanol. A one-carbon atom alcohol made from natural gas, coal, or biomass. Also called methyl or wood alcohol.

Miscible flooding. An enhanced recovery method in which a fluid, usually carbon dioxide, is injected into a well and dissolves in the oil which then flows more easily to producing wells. Miscible substances will mix together to form a homogeneous mixture.

Mmcf. Million cubic feet. The cubic foot is a standard unit of measure for quantities of gas at the atmosphere pressure.

Moratorium. A formally announced suspension of a given type of activity, which can come either at the initiative of the organization(s) concerned or through the intervention of a legal authority.

National Ambient Air Quality Standards. Standards in the Clean Air Act that set maximum concentrations that should be allowed nationwide for air pollutants. Standards have been set for seven pollutants: carbon monoxide, hydrocarbons, lead, nitrogen dioxide, ozone, particulates, and sulfur dioxide.

Natural gas. A mixture of hydrocarbon compounds and small amounts of various nonhydrocarbons (such as carbon dioxide, helium, hydrogen sulfide, and nitrogen) existing in the gaseous phase or in solution with crude oil in natural underground reservoirs.

Natural gas hydrates. Ice-like mixtures of methane and water, sometimes found in permafrost or in sediments beneath the ocean floor.

Natural gas liquids (NGL). Portions of natural gas that are liquefied at the surface in lease separators, field facilities, or gas processing plants, leaving dry natural gas. They include, but are not limited to, ethane, propane, butane, natural gasoline, and condensate.

Nonattainment areas. Regions of the country that do not meet the National Ambient Air Quality Standards of the Clean Air Act for one or more of the seven regulated air pollutants.

Octane number. A rating used to grade the relative anti-knock properties of various gasolines. A high octane fuel has better anti-knock properties than one with a low number.

Offset policy. Policy that new industrial plants could not be built in nonattainment areas as defined by the Clean Air

Act unless pollution from existing factories was reduced enough to compensate for the pollution expected from the new plant. Thus a company wanting to build in a nonattainment area must keep its own emissions down in that area or buy emissions offsets from other companies.

Offshore platform. A fixed structure from which wells are drilled offshore for the production of oil and natural gas.

Oil shale. A fine-grained, sedimentary rock that contains a solid substance, kerogen, which is partially formed oil. Kerogen can be extracted in the form of shale oil by heating the shale.

OPEC. The Organization of Petroleum Exporting Countries, an international oil cartel that includes Saudi Arabia, Kuwait, Iran, Iraq, Venezuela, Qatar, Libya, Indonesia, United Arab Emirates, Algeria, Nigeria, Ecuador, and Gabon.

Outer Continental Shelf (OCS). A gently sloping underwater plain that extends seaward from the coast. Technically, it includes only those lands between the end of state jurisdiction (three miles in most areas, but about ten miles for Texas and parts of Florida) and the 200-meter water depth. However, the term OCS is now used by the government and the petroleum industry to include both the continental shelf and the continental slope to the 2,500-meter water depth.

Overthrust belt. A geological system of faults and basins in which geologic forces have thrust layers of older rock above strata of newer rock that might contain oil or natural gas. The Eastern Overthrust Belt runs from eastern Canada through Appalachia into Alabama. The Western

Overthrust Belt runs from Alaska through western Canada and the Rocky Mountains into Central America.

Particulates. Tiny particles of soot, ash, and other solids or liquids that are emitted into the air.

Passive solar. Architectural designs in buildings that take advantage of site and building materials to enhance the amount of solar radiation turned into useful interior heat during cold periods and to minimize absorption of solar heat during warm periods.

Permeability. A measure of the capacity of a rock or stratem to allow water or other fluids such as oil to pass through it. (See porosity.)

Petrochemicals. Chemicals derived from crude oil or natural gas, include ammonia, carbon black, and thousands of organic chemicals.

Petroleum. Strictly speaking, crude oil. In a broader sense, it refers to all hydrocarbons, including oil, natural gas, natural gas liquids, and related products.

Photovoltaics. The conversion of sunlight directly into electricity by means of solar cells.

Plutonium. A radioactive element that can be a raw material in the manufacture of nuclear weapons or a waste product of processes yielding atomic energy.

Porosity. A measure of the amount of void space or pores within rock that affects the amount of liquids and gases (such as crude oil and natural gas) that rock can contain.

Prevention of Significant Deterioration (PSD) areas. Areas that meet or exceed Clean Air Act standards for specific pollutants. Emissions in these areas are regulated so that the air quality does not deteriorate beyond certain levels.

Price control. Setting limits on prices (usually maximum limits) by government order.

Primary recovery. Extracting oil from a well by allowing only the natural water or gas pressure in the reservoir to force the petroleum to come to the surface, without pumping or other assistance. Also called flush production.

Production. A term commonly used for natural resources actually taken out of the ground.

Proved reserves. An estimate of the amount of oil or natural gas believed to be recoverable from known reservoirs under existing economic and operating conditions.

Public lands. Any land or land interest owned by the federal government within the fifty states. The term does not include offshore federal lands or lands held in trust for Indians and other native American groups.

Pyrolysis. Application of heat to pulverized coal in the absence of air to break the coal molecules into liquids and gases.

R and D. Research and development.

Reclamation. Restoring land to its original condition by regrading contours and replanting after the land has been mined, drilled, or otherwise undergone alteration from its original state.

Recoverable resources. An estimate of resources, including oil and/or natural gas, both proved and undiscovered, that would be economically extractable under specified price-cost relationships and certain technological conditions.

Rem. A measure of radiation exposure, specifically the dosage of radiation that will cause the same amount of biological injury to human tissue as one roentgen of x-ray or gamma ray dosage. Usually used in fractional amounts, such as millirems (1 millirem = 0.001 rem).

Rental. The amount periodically paid by a leaseholder to a landowner for the right to use property for purposes set out in the lease.

Reserve. A porous and permeable underground formation of producible oil and/or natural gas, confined by impermeable rock or water barriers and characterized by a single natural pressure system.

Reserve-production ratio. A way of showing how many years it would take to use up the nation's proved reserves of oil and natural gas at current production levels. Such estimates do not include oil or gas that may be discovered in the future or resources known to exist but now regarded as uneconomic to produce.

Residual fuel. Heavy oil used by utilities and industries for fuel.

Retort. Any closed vessel or facility for heating a material to cause a chemical reaction.

Retorting. Any of a variety of methods by which a carbonaceous material is heated, generally above 700 degrees Fahrenheit, to decompose the material into gases, oils, tars, and carbon. Heat for retorting is obtained by burning a portion of the raw feed and/or fuels derived from the process. Oil shale is processed by retorting.

Retrofitting. See backfitting.

Return on investment (rate of return). A measure of the profitability of a business enterprise. The general form is profit divided by investment, but the calculations can take many alternative forms. Investment can refer to stockholders' equity only or can also include long-term borrowed funds for all resources available to a company, that is total assets. Profits can be measured before or after corporate income taxes.

Royalty. A payment to a landowner or mineral rights owner (which can be a government body or private party) by a leaseholder on each unit of resource produced. Oil and natural gas royalties are usually paid in cash, as a percentage of the value of production. However, in some cases, the landowner or mineral rights owner receives a percentage of the actual petroleum produced.

SASOL. South African Synthetic Oil Limited. A government-owned synthetic fuel facility located in Secunda, South Africa, where South African coal is gasified and a broad range of synthetic fuels are produced.

Scrubber. A pollution control device in the stack of a coal-burning facility which uses liquid spray to remove pollutants, especially sulfur dioxide, from emissions. The process is called flue gas desulfurization.

Sedimentary rock. Rock formed of sediments, usually deposited in a marine environment, such as shale or sandstone. Petroleum is found in sedimentary rock.

Seismic exploration. A method of prospecting for oil or gas by sending shock waves into the earth. Reflections of the shock waves that bounce off rock strata are recorded on magnetic tape. The time it takes for the wave to return to the surface can be interpreted by experts to indicate the depth of specific strata and the composition of intermediate strata.

Service well. A well drilled in a known oil or natural gas field to inject liquids to enhance recovery, dispose of salt water, or for purposes other than actual production.

Severance tax. A tax paid to a state by producers of mineral resources (including oil and gas) in the state. The tax is usually levied as a percent of the value of the oil or gas severed from the earth. It may also be expressed as a specific amount per barrel of oil or thousand cubic feet of gas.

Shale oil. The hydrocarbon substance produced from the decomposition of kerogen when oil shale is heated in an oxygen-free environment. Raw shale oil resembles a heavy, viscous, low-sulfur crude but can be upgraded to produce a good quality sweet crude.

Sludge. In discussions of environmental controls, the mud-like residue that results from the cleaning process of scrubbers or certain other devices designed to prevent solid particulates from entering the environment.

Solar thermal energy. The use of the sun's heat for space heating (or through refrigeration coils for cooling).

Solvent refined coal (SRC). A coal extract derived through the use of solvents. Crushed coal is mixed in a solvent at high temperature and pressure in the presence of hydrogen. The process produces an ash and a sulfur free solid (SRC-I) or liquid (SRC-II). After processing, the ash is removed from the solvent and the solvent recycled.

Sour crude oil. Crude oil that contains significant amounts of hydrogen sulfide. Often less valuable than sweet crude oil.

Spent shale. Shale that is left over after kerogen (shale oil) has been removed.

State OCS jurisdiction. The area where coastal states own mineral rights on offshore lands. In general, the boundary is three nautical miles out from shore, except off

Texas and the Gulf Coast of Florida where the states own three leagues (about ten nautical miles) out from shore.

Steam coal. A quality of coal used by utilities for generating steam to make electricity. Usually has a lower heat content than metallurgical coal.

Stocks. Inventories. Widely used in the petroleum industry to designate inventories of crude oil and other oil products at refineries, bulk terminals, and in pipelines.

Strata. Layers of rock.

Strategic Petroleum Reserve (SPR). A stockpile of oil maintained by the United States. The United States government purchases crude oil and pumps it into underground salt domes for use in case of an import supply interruption, shortage, or other emergency.

Stratigraphic test. A well drilled specifically to obtain detailed information on the composition of a rock formation that might lead to the discovery of oil or natural gas.

Stripper gas well. A well that produces an average of less than 60,000 cubic feet a day, measured over a ninety-day period.

Stripper oil well. A well capable of producing no more than ten barrels of oil a day. Nearly three-fourths of the oil wells in the United States are strippers. Together they account for one out of every seven barrels of oil produced.

Subbituminous coal. Coal ranked between bituminous and lignite in heat value and overall quality. It has a range of heat values between 8,300 and 11,500 Btu per pound.

Sulfur dioxide emissions. Heavy, pungent, toxic gases released when fuel containing sulfur is burned.

Sweet crude oil. Crude oil that is low in sulfur. Often more valuable than sour crude oil.

Synthetic crude oil (syncrude). A crude oil derived from processing a carbonaceous material. Oil extracted from shale or unrefined oil from coal conversion plants are syncrudes.

Synthetic gas. Gas made from solid hydrocarbons such as coal, oil shale, or tar sands.

Synthetic fuels (synfuels). Fuels that are produced through complex chemical conversions of such natural fossil substances as coal and oil shale. Synthetic fuels are comparable in chemical structure and energy value to oil products and natural gas.

Tanker. Oceangoing ship specially designed for carrying crude oil and other liquid petroleum products.

Tar sands. Rocks containing highly viscous hydrocarbons (other than coal or oil shale) that are not recoverable by primary production methods. That is, the hydrocarbons in tar sands cannot readily move as a fluid under their own reservoir energy. The hydrocarbons in heavy oil sands, in contrast, will flow slowly under their own reservoir energy.

Tcf. Trillion cubic feet. The cubic foot is a standard unit of measure for quantities of gas at atmospheric pressure.

Therm. A measure of heat content. One therm equals 100,000 British thermal units.

Thermal recovery. An enhanced recovery method using heat to thin oils that are too thick to flow to producing wells. One method uses the heat created by injecting oxygen and starting a fire in the well. Another consists of injecting steam.

Tight reservoirs. Soil or rock formations with low permeability to oil, water, or natural gas.

Trans–Alaska Pipeline. A crude oil pipeline from the Northern Slope of Alaska above the Arctic Circle to the ice-free port of Valdez in southern Alaska.

Undiscovered recoverable resources. Resources outside of known fields, estimated from broad geologic knowledge and theory.

Unit train. A train of at least one hundred cars specially designed to haul and unload coal.

Viscosity. The measure of a liquid's internal friction or of its resistance to flow.

Waterflooding. The most common enhanced recovery method (usually referred to as secondary recovery). Water is pumped into an oil reservoir to push the oil toward producing wells. This method of recovery is often used after a field's own internal pressure is no longer sufficient to provide adequate oil production.

Water quality standards. Standards adopted by states under the Clean Water Act for quantities of pollution allowed in water bodies. Water quality standards are not the major method of pollution control under the Clean Water Act, which primarily designates required control technology rather than particular standards of water quality.

Wetlands. Lands with very moist soil, such as tidal flats or swamps.

Wildcatter. An operator who drills the first well in unproven territory.

Wilderness land. Land withdrawn from development by Congress in order to preserve its pristine characteristics as set forth in the National Wilderness Preservation System Act of 1964.

Windfall Profits Tax. A U.S. tax on crude oil production (actually an excise tax rather than a tax on profits). The tax is a specified percentage of the difference between the sales price in the field and a base price established by the government for each of various oil categories. The percentage and the base price both vary depending on how a given field had been treated under price control regulations. The base prices are automatically increased each year. Repeal of the tax was signed into law as part of the massive trade bill by President Ronald Reagan on August 24, 1988.

Withdrawal of government lands. Restricting the use of government lands by holding them for specific public purposes. Such restrictions virtually always include prohibition of exploration for and production of oil, natural gas, or other minerals.

SCHOOLS SPECIALIZING IN PETROLEUM CURRICULUMS

A number of colleges, universities, and other accredited learning institutions offer a wide range of petroleum curriculums, from the traditional earth sciences to the newer degrees, such as petroleum land management. The schools listed here should provide an excellent education in many of the earth sciences. For more information, contact the admissions department or a specific department at the university, such as petroleum engineering department or geology department.

For a more detailed listing, check with the Society of Petroleum Engineers at the address listed in appendix C. This organization publishes a resource document listing petroleum engineering and technology schools. Listed in the publication is the school, address, name and phone number of a contact person, faculty and their major fields of interest, curriculum description, and program admission requirements. The book also lists the number of under-graduates and graduates enrolled in the curriculum and number of degrees conferred. For the specific year, the names of student officers and candidates for degrees are also listed.

Alabama

University of Alabama
Tuscaloosa, AL 35487

Alaska

University of Alaska
Fairbanks, AK 77701

California

Stanford University
Stanford, CA 94305

University of California
Berkeley, CA 94720

University of Southern California
University Park
Los Angeles, CA 90007

Colorado

Colorado School of Mines
Golden, CO 80401

Kansas

University of Kansas
Lawrence, KS 66045

Louisiana

Louisiana State University
Baton Rouge, LA 70803

Louisiana Tech University
Ruston, LA 71272

University of Southwestern Louisiana
Lafayette, LA 70504

Mississippi
Mississippi State University
Mississippi State, MS 39762

Missouri
University of Missouri at Rolla
Rolla, MO 65401

Montana
Montana College of Mineral Science & Technology
Butte, MT 59701

New Mexico
New Mexico Institute of Mining & Technology
Campus Station
Socorro, NM 87801

Ohio
Marietta College
Marietta, OH 45750

Oklahoma
University of Oklahoma
Norman, OK 73019

University of Tulsa
Tulsa, OK 74104

Pennsylvania
Pennsylvania State University
University Park, PA 16802

Texas
Texas A&I University
Kingsville, TX 78363

Texas A&M University
College Station, TX 77843

Texas Tech University
Tech Station
Lubbock, TX 79409

The University of Texas
Austin, TX 78712

West Virginia
West Virginia University
Morgantown, WV 26506

Wyoming
University of Wyoming
University Station
Laramie, WY 82071

INTERNATIONAL AND NATIONAL ORGANIZATIONS AND ASSOCIATIONS

The following organizations and associations may be contacted for more information on the petroleum industry.

American Association of Petroleum Geologists
P.O. Box 979
Tulsa, OK 74101
(918) 584-2555

American Chemical Society
Education Division, Career Services
1155 16th St., NW
Washington, DC 20036
(202) 872-4450

American Geological Institute
4220 King St.
Alexandria, VA 22302
(703) 379-2480

American Geophysical Union
  2000 Florida Ave., NW
  Washington, DC 20009
  (202) 462-6903

American Institute of Chemical Engineers
  345 East 47th St.
  New York, NY 10017
  (212) 705-7660

American Institute of Chemists
  7315 Wisconsin Ave., Suite 525–E
  Bethesda, MD 20814
  (301) 652-2447

American Petroleum Institute (API)
  Public Relations Department
  1220 L Street, NW
  Washington, DC 20005
  (202) 682-8000

Association of Oilwell Servicing Contractors
  6060 N. Central Expressway, Suite 538
  Dallas, TX 75206
  (214) 692-0771

Energy Information Administration
  Forrestal Building, Room 1F–048
  1000 Independence Ave., SW
  Washington, DC 20585
  (202) 586-5000

Independent Petroleum Association of America (IPAA)
  1101 16th St. NW
  Washington, DC 20036
  (202) 857-4777

International Association of Drilling Contractors (IADC)
　P.O. Box 4287
　Houston, TX 77210

Marine Technology Society
　2000 Florida Ave., NW, Suite 500
　Washington, DC 20009
　(202) 462-7557

National Conference of Chemical Technician Affiliates
　American Chemical Society
　Ms. Naomi Williams, President
　Monsanto Company, Q3D
　800 North Lindbergh Blvd.
　St. Louis, MO 63167

Society of Exploration Geophysicists
　P.O. Box 70240
　Tulsa, OK 74170

Society of Petroleum Engineers (SPE)
　Publications Department
　P.O. Box 833836
　Richardson, TX 75083-3836
　(214) 669-3377

BIBLIOGRAPHY

The following sources were used to compile information for this book. The books, publications, and other material listed below are excellent sources for further study of the petroleum industry and the many opportunities available.

BOOKS

Anderson, Kenneth E. and Bill D. Berger. *Modern Petroleum: A Basic Primer of the Industry.* 2nd ed. Tulsa, Okla.: PennWell Books, PennWell Publishing Co., 1981.

Baker, Ron. *A Primer of Oilwell Drilling.* 4th ed. Austin, Tex.: Petroleum Extension Service, The University of Texas at Austin, in cooperation with International Association of Drilling Contractors, Houston, Tex., 1979.

Clark, James A. *The Chronological History of the Petroleum and Natural Gas Industries.* Edited by Charles A. Warner and research directed by Harry E. Walton. Houston, Tex.: Clark Book Co. Inc., 1963.

Knowles, Ruth Sheldon. *The First Pictorial History of the American Oil and Gas Industry 1859–1983*. Athens, Ohio: Ohio University Press, 1983.

Sampson, Anthony. *The Seven Sisters: The Great Oil Companies and the World They Made*. Coronet Edition 1976. London: Coronet Books, Hodder and Stoughton Ltd., 1975.

Wheeler, Robert R. and Maurine Whited. *Oil: From Prospect to Pipeline*. 4th ed. Houston, Tex.: Gulf Publishing Co., November 1981.

PUBLICATIONS

The ABC's of Oil. A Scriptographic Booklet by Channing L. Bete Co. Inc., South Deerfield, Mass., 1982.

About Careers in the Petroleum Industry. A Scriptographic Booklet by Channing L. Bete Co. Inc., South Deerfield, Mass., 1983.

Energy Security White Paper: U.S. Decisions and Global Trends. American Petroleum Institute: Washington, D.C., November 1988.

Facts About Oil. American Petroleum Institute, Washington, D.C.

Fundamentals of Petroleum. 3rd ed. Petroleum Extension Service, The University of Texas at Austin, Austin, Tex., 1986.

Occupational Outlook Handbook. 1988–89 ed. Washington, D.C.: U.S. Department of Labor, Bureau of Labor Statistics, 1988.

Oil from the Earth. Rev. ed. 1983. Society of Petroleum Engineers of AIME. Richardson, Tex., 1974.

Oil in Depth: The Story of Petroleum Exploration and Production. Amoco, Chicago.

The Oil & Gas Producing Industry in Your State. Petroleum Independent, Independent Petroleum Association of America, Washington, D.C., September 1988.

Petroleum Engineering and Technology Schools 1987–88. Society of Petroleum Engineers of AIME, Richardson, Tex. 1987.

Petroleum Exploration: Continuing Need. American Petroleum Institute, Washington, D.C.

VGM CAREER BOOKS

OPPORTUNITIES IN

*Available in both
paperback and hardbound
editions*

Accounting Careers
Acting Careers
Advertising Careers
Agriculture Careers
Airline Careers
Animal and Pet Care
Appraising Valuation Science
Architecture
Automotive Service
Banking
Beauty Culture
Biological Sciences
Biotechnology Careers
Book Publishing Careers
Broadcasting Careers
Building Construction Trades
Business Communication Careers
Business Management
Cable Television
Carpentry Careers
Chemical Engineering
Chemistry Careers
Child Care Careers
Chiropractic Health Care
Civil Engineering Careers
Commercial Art and Graphic
 Design
Computer Aided Design
 and Computer Aided Mfg.
Computer Maintenance Careers
Computer Science Careers
Counseling & Development
Crafts Careers
Dance
Data Processing Careers
Dental Care
Drafting Careers
Electrical Trades
Electronic and Electrical
 Engineering
Energy Careers
Engineering Technology Careers
Environmental Careers
Fashion Careers
Fast Food Careers
Federal Government Careers
Film Careers
Financial Careers
Fire Protection Services
Fitness Careers
Food Services
Foreign Language Careers
Forestry Careers
Gerontology Careers
Government Service
Graphic Communications
Health and
 Medical Careers
High Tech Careers
Home Economics Careers
Hospital Administration
Hotel & Motel Management
Human Resources Management
 Careers

Industrial Design
Insurance Careers
Interior Design
International Business
Journalism Careers
Landscape Architecture
Laser Technology
Law Careers
Law Enforcement and
 Criminal Justice
Library and Information
 Science
Machine Trades
Magazine Publishing Careers
Management
Marine & Maritime Careers
Marketing Careers
Materials Science
Mechanical Engineering
Medical Technology Careers
Microelectronics
Military Careers
Modeling Careers
Music Careers
Newspaper Publishing
 Careers
Nursing Careers
Nutrition Careers
Occupational Therapy
 Careers
Office Occupations
Opticianry
Optometry
Packaging Science
Paralegal Careers
Paramedical Careers
Part-time & Summer Jobs
Petroleum Careers
Pharmacy Careers
Photography
Physical Therapy Careers
Plumbing & Pipe Fitting
Podiatric Medicine
Printing Careers
Property Management
 Careers
Psychiatry
Psychology
Public Health Careers
Public Relations Careers
Purchasing Careers
Real Estate
Recreation and Leisure
Refrigeration and Air
 Conditioning Trades
Religious Service
Restaurant Careers
Retailing
Robotics Careers
Sales Careers
Sales & Marketing
Secretarial Careers
Securities Industry
Social Science Careers
Social Work Careers
Speech-Language Pathology
 Careers
Sports & Athletics
Sports Medicine

State and Local Government
Teaching Careers
Technical Communications
Telecommunications
Television and Video Careers
Theatrical Design
 & Production
Transportation Careers
Travel Careers
Veterinary Medicine Careers
Vocational and Technical
 Careers
Word Processing
Writing Careers
Your Own Service Business

CAREERS IN

Accounting
Business
Communications
Computers
Education
Engineering
Health Care
Science

CAREER DIRECTORIES

Careers Encyclopedia
Occupational Outlook Handbook

CAREER PLANNING

Handbook of Business and
 Management Careers
Handbook of Scientific and
 Technical Careers
How to Get and Get Ahead
 On Your First Job
How to Get People to Do
 Things Your Way
How to Have a Winning
 Job Interview
How to Land a Better Job
How to Prepare for College
How to Run Your Own Home
 Business
How to Write a Winning
 Résumé
Joyce Lain Kennedy's Career Book
Life Plan
Planning Your Career Change
Planning Your Career of
 Tomorrow
Planning Your College
 Education
Planning Your Military Career
Planning Your Young Child's
 Education

SURVIVAL GUIDES

High School Survival Guide
College Survival Guide

VGM Career Horizons
a division of *NTC Publishing Group*
4255 West Touhy Avenue
Lincolnwood, Illinois 60646 1975